Frequency Synthesis:
Techniques and Applications

Frequency Synthesis: Techniques and Applications

Edited by
Jerzy Gorski-Popiel
Staff Member
M.I.T. Lincoln Laboratory

Contributing Authors:
Jerzy Gorski-Popiel, Staff Member, M.I.T. Lincoln Laboratory
Thomas S. Seay, Associate Group Leader, M.I.T. Lincoln Laboratory
Ben H. Hutchinson, Jr., Associate Group Leader, M.I.T. Lincoln Laboratory
Carl H. Gundel, Senior Engineering Specialist, GTE Sylvania
Joseph Tierney, Staff Member, M.I.T. Lincoln Laboratory
George H. Lohrer, Engineering Staff, General Radio Corporation

A volume published under the sponsorship of
the IEEE Educational Activities Board

IEEE PRESS

The Institute of Electrical and Electronics Engineers, Inc. New York

Copyright © 1975 by
THE INSTITUTE OF ELECTRICAL AND ELECTRONICS ENGINEERS, INC.
345 East 47 Street, New York, N.Y. 10017
All right reserved.

International Standard Book Numbers: Clothbound: 0-87942-039-1

Library of Congress Catalog Card Number 74-82502

PRINTED IN THE UNITED STATES OF AMERICA

Preface

There is a genuine need for state-of-the-art texts that combine, in one volume, a great deal of information about a single specialized subject. Such books would substantially increase the efficiency of an engineer faced with the need to start a new project in an area not familiar to him. This book is intended to satisfy this need in the area of frequency synthesis. Thus the intended readers range from recent electrical engineering graduates to long-time practicing engineers.

The term "frequency synthesizer" has been used to describe any device generating an output frequency from one or more input frequencies. However, in contemporary usage a frequency synthesizer is a device whose output frequency, f_o, is intended to be a rational multiple of a single standard frequency, f_s. Thus on the average, $f_o = (N/M)f_s$, where M and N are integers. The synthesizers of interest here can generate one out of a large number of output frequencies on external command. However, the techniques are applicable to generators of a fixed output frequency.

The first description in the literature of such a device occurred during World War II (Finden, 1943). During the following 30 years many different systems for frequency synthesis have been developed. These systems may be broadly divided into direct techniques, indirect techniques, and all-digital techniques.

In direct frequency synthesis, the output frequency is obtained directly from the standard frequency by the operations of mixing, filtering, multiplication, and division. Although many of the earliest synthesizers used direct synthesis, the introduction of solid state technology permitted widespread application of this method.

In contrast, indirect frequency synthesizers derive the output frequency from a secondary oscillator that is usually phase locked (sometimes only frequency locked) to the primary frequency reference. Strictly speaking, such systems are not "synthesizers," although they are commonly described as such. They have also been referred to by other names, such as stabilized master oscillators (SMOs), but such terminology is neither consistent nor universally employed. Although early indirect frequency synthesizers used vacuum tube technology, the introduction of high-speed integrated circuits greatly facilitated utilization of this method.

With the advent of medium and large scale integration of digital circuits, the third approach — digital frequency synthesis — became practical. In this approach, samples of the output sine wave are generated digitally and then converted by digital-to-analog conversion and filtering to the desired output sinusoid.

Although the impact of frequency synthesizers has been great, it is the opinion of the authors that frequency synthesizers should become much more widely used in the next few years. Thus the purpose of this book is to stimulate the widespread application of a relatively mature technology by:

1. Listing the basic properties of synthesizers as a class and describing a range of applications, most of which were impractical before the introduction of synthesizers (Chapter 1).

2. Comparing the direct, indirect, and all digital approaches to frequency synthesis on the basis of achievable performance and complexity (Chapter 2).

3. Presenting case studies of the three different techniques in sufficient detail to permit the reader to be comfortable in the application and design of synthesizers (Chapters 3, 4 and 5).

4. Discussing the design and measurement problems of synthesizer noise and alternate synthesizer configurations (Chapter 6).

Detail considered sufficient to understand the basic principles and limitations of each approach is included. The references will guide the reader to a more penetrating comprehension of the subject even though many of the hundreds of articles on frequency synthesizers published during the last 30 years have not been included. Also, many of the commercial synthesizer manufacturers publish application notes. Lists of commercial manufacturers can be found in the current issues of Electronics Buyers Guide, Electronic Engineers' Master, and in an article surveying commercial instruments by Runyon in 1973.

The chapters of this book were presented initially at IEEE seminars held in Boston and New York during February and March of 1972, and subsequently updated. The presentations and book format are intended to facilitate rapid and accurate dissemination of timely information. The cooperation and contributions of M.I.T. Lincoln Laboratory, Sylvania Electronic Systems, General Radio Company, and the IEEE are gratefully acknowledged.

Jerzy Gorski-Popiel

Contents

Chapter I
Applications of Frequency Synthesizers
T. S. Seay

Many different properties of synthesizers must be characterized to determine performance in a particular application. The first part of this chapter considers these various properties, and in some cases, describes measurement techniques. Typical parameter values are indicated. The remainder of the chapter addresses specific applications.

The broad range of frequency synthesizer applications may be arbitrarily divided into three classes:

   a. Channelized communications, which supplied much of the early impetus for the development of frequency synthesizers.

   b. Measurements, where many more measurements are performed for a given cost.

   c. Advanced systems that are impractical without remote programmability and rapid switching speed.

The requirements of the first two application groups could, in principle, be satisfied by a high-quality signal generator and frequency counter, but usually at much greater expense.

A. SYNTHESIZER PROPERTIES

The frequency synthesizer is a relatively complicated system involving many components, and usually a number of properties must be specified to characterize performance. But there is no uniform standard to describe or measure many of these parameters. This may be reasonable as appropriate parameters often depend on the application.

1. Frequency Standard

The greatest advantage of a frequency synthesizer is that the output frequency is essentially a rational multiple of the input frequency. Although the average output frequency is an exact rational multiple of the input frequency, for short time periods (short-term stability) the output frequency may suffer phase perturbations from internal circuit noise. Amplitude perturbations also occur, but these are usually much smaller. The reference frequency can be chosen for convenience of oscillator and crystal design. Since there is only one reference oscillator, it can provide, by design, high short- and long-term stability with low noise. Hence, considerable design freedom is permitted because long-term output frequency stability is not dependent on the frequency stability of synthesizer components.

Most commercial frequency synthesizers include an internal frequency standard. Many also provide an external input line to permit use of a higher quality external standard. Low cost (up to several $100) crystal oscillators can be purchased with a

1

long-term aging rate of about 1 in 10^6 parts per year with comparable compensation for drift over a $-55°$ to $+100°$C temperature range. Higher quality commercial crystal oscillators provide an aging rate of less than 5 in 10^{10} parts per day, with less than 3 in 10^9 parts variation in output frequency over a $0°$ to $+50°$C temperature range.

2. Resolution

Resolution is the minimum frequency difference between any two adjacent output frequencies. Synthesizers usually generate all frequencies within a specified output band with identical spacing as there is little economic advantage to delete some frequencies or to provide nonuniform spacing. Resolution depends largely on the application, and may be as crude as 100 kHz for general communications, or as fine as 0.01 Hz or better for low-rate tracking systems or highly accurate frequency translations. Many commercial synthesizers provide a continuously adjustable interpolation oscillator for virtually infinite resolution but with degraded stability. However, the incremental cost of high resolution by synthesis is almost negligible in many contemporary designs.

3. Number of Frequencies

The economic advantage of most frequency synthesizers results from the large number of high performance oscillators that are eliminated. Useful synthesizers may be able to generate an output on any one of as few as 100 discrete output frequencies, although some test instruments can generate any one of 5×10^9 discrete output frequencies.

4. Programmability

The output frequencies of essentially all modern frequency synthesizers are specified by the DC levels on a set of control wires. The advantages of this approach are:

    a.   Remote control.

    b.   Optimum layout to minimize spurious radiations, even when frequency control is derived from the front panel.

    c.   More rapid and reliable control than with mechanical systems.

Where the frequency command rate is not high, the trend is to use a serial data transfer line to provide frequency control information. This approach minimizes the number of cable connections to the synthesizer, thereby reducing one of the most persistent sources of unreliability in complicated systems.

5. Switching Time

The switching time, i.e., the time between a command to a different frequency is given and the time when the output is useful, is an important parameter for applications requiring frequent frequency switching. Some applications require that the transition be phase continuous. The usual switching speed quotation is for the worst case.

However, there are designs where switching speed is almost proportional to the difference between the initial and final frequencies; for others, switching speed is determined by the number of frequency control bits that must be changed. Thus where switching speed is critical, switching requirements must be specified in detail. Typical worst-case switching speeds range from 1 msec to about 10 μsec, and synthesizers with switching speeds of less than 1 μsec have been built.

Another important parameter for some applications is the rate at which successive commands to change frequency can be accepted by the synthesizer. The maximum command period is roughly the synthesizer switching time when parallel control is used. In serial control, the maximum command rate is typically 10 to 50 percent less than $1/N$ times the control word shift rate, where N is the number of bits in a control word.

6. Spurious Signals and Noise

As pointed out, a frequency synthesizer's output is not a perfect sinusoid. Corruption of the desired output signal, usually referred to as the "carrier," may be divided into (a) coherent spurious signals generated in various nonlinear operations such as mixing, and (b) noncoherent noise outputs due to internal circuit noises. Because synthesizer output is normally amplitude leveled to a very high degree of accuracy, the primary output noise is phase noise.

The best way to specify the spurious emission and/or noise level depends on the application. The usual procedure is to specify the maximum power within a specified bandwidth in dB relative to the carrier power for all bandwidths offset from the carrier by a specified minimum frequency separation. Thus a common test is to measure the total power from a filter of specified characteristics, when the center frequency of the measuring filter is offset from the output carrier by some minimum frequency separation. A spectrum analyzer is often used for this purpose.

Measurements of spurious signals and noise very close to the carrier have been relatively difficult in the past. However, by using a phase-lock loop to null out the carrier signal in a balanced mixer (Fig. 1) spurious signals down to −120 dB can be detected at very small frequency offsets from the carrier (Horn, 1969).

Typical spurious/noise emission requirements range from −50 dB relative to the carrier measured in 100-kHz bandwidth, to −140 dB relative to the carrier measured in a 3-kHz bandwidth. Typical performance is for spurious/noise power −80 dB relative to the carrier in a 25-kHz bandwidth offset 50 kHz or more. Various synthesizers have been reported with all nonharmonic discrete spurious emissions better than 100 dB below the carrier, and with phase noise below −130 dB/Hz relative to the carrier for frequencies greater than 1 kHz from the carrier.

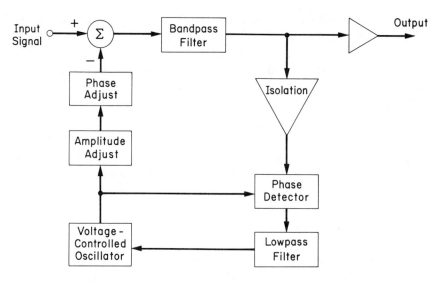

Fig. 1. Measurement of spurious signals and noise close to the carrier.

7. Short-Term Stability

Short-term stability refers to fluctuations in the output frequency on a time scale typically less than 1 second. Specifically, the instantaneous output voltage may be written as

$$V(t) = V \sin(2\pi f_o t + 2\pi \Delta f t) \quad .$$

Then the fractional frequency deviation from nominal is $\Delta f/f_o$. There are two proposed definitions for the measure of frequency stability (Barnes, et al., 1971):

a. In the frequency domain, specify the one-sided power spectral density of the process $\Delta f/f_o$.

b. In the time domain, let $\bar{y}_k$ be the average value of $\Delta f/f_o$ over the time interval $(t_k, t_k + \tau)$. τ is called the averaging interval. Since this average value fluctuates with time, the second measure of stability is the variance:

$$\sigma_y^2(\tau) = \left\langle \frac{(\bar{y}_{K+1} + \bar{y}_K)^2}{2} \right\rangle \quad ,$$

where $\langle\;\rangle$ denotes infinite time average.

In practice, the spectrum and the variance can only be estimated. Statistical confidence limits for the estimates should be given.

For very small averaging times, τ, or for large frequencies in the power spectrum, short-term instabilities result from spurious signals and noise (previous section). In general, short-term instabilities are attributed to both $1/f$ and additive gaussian noise of phase and frequency.

A comprehensive discussion of the ways in which frequency stability and the associated measurements can be characterized, as well as the interrelations among the frequency-domain and time-domain stability definitions, has been given by Barnes, et al., 1971. Other discussions are given by Tykulski and Maldewin, 1967; Shields, 1969; Leeson, 1971; Reynolds, 1972.

B. SYNTHESIZERS FOR CHANNELIZED COMMUNICATIONS

Probably the greatest impetus to frequency synthesizer development has been the demand for communication capability within a limited spectrum. By providing accurate frequency control at tractable costs, frequency synthesizers provide a maximum number of channels for an available total bandwidth and modulation technique. Perhaps equally important, this frequency control can be obtained with relatively untrained operators.

1. HF Communications

One of the earliest widespread applications of frequency synthesizers was in HF communications for fixed and shipborne installations. In addition to permitting relatively close channel spacing by eliminating guard bands to accommodate frequency drift, frequency synthesizers with accurate frequency standards also permitted single sideband without a tracking carrier for point-to-point communications. Elimination of tracking carriers reduced interference between signals on close or common frequency channels (Craiglow and Martin, 1956). Furthermore, it became possible to build equipment with rapid, accurate tuning to any channel in the HF band, permitting both quick adaptation to varying propagation and interference conditions, and accurate channel surveillance to permit rapid linking without extensive calling and tuning (Young and Johnson, 1957).

Typical transmitter frequency resolution requirements range from 100 Hz to 1 kHz. Some receivers may require up to 1-Hz resolution (East, 1962). Estimates of the required frequency accuracy for transmitting range from ± 0.5 in 10^6 (Craiglow and Martin, 1956; East, 1962) to 1 in 10^8 (Rozov, 1958) for a single sideband without carrier. It was estimated recently that a frequency accuracy of about 1 in 10^7 is necessary for future, single sideband, secure voice, tactical applications (Layden, 1969).

Spurious signal requirements on HF equipments are usually quite severe because of the likely collocation of high power transmitters and sensitive receivers. Typical transmitter requirements limit spurious products to at least 80 dB below the nominal output and often as much as 120 dB (Flicker and Gerhold, 1967).

The evolution of HF radio receiver design is interesting. Early receivers used continuously tunable local oscillators, which were replaced by frequency synthesizers during the last 20 years. A recent design (Peterson, 1971; Peterson, 1972) uses a shaft encoder output to drive a digital frequency synthesizer with 100-Hz resolution. Tuning by the shaft encoder knob provides the "feel" of a continuously tuning receiver, combined with the frequency stability of a frequency synthesizer.

Ionospheric sounding — transmitting HF radio pulses vertically from the ground and receiving the totally reflected signal at the same location — has been developed to not only enhance the reliability of HF communications, but also to obtain environmental data and for experiments in plasma physics and wave propagation. The resulting plot of travel time (interpreted as a virtual height of reflection) versus frequency is called a swept-frequency ionogram. An automatic ionospheric sounding system, using a frequency synthesizer for automatic tuning, is expected to contribute significantly to understanding the ionosphere (Wright, 1969). Another such system using high-resolution array antennas steerable in azimuth and elevation provides one-way propagation and two-way (backscatter) measurements to monitor traveling ionospheric disturbances, skip distances, plasma frequencies, and possibly, the state of the sea over a large remote area (Tveten and Hunsucker, 1969; Barrick, 1973).

2.   Mobile Communications

Restrictions on weight, size, power, cost and (especially for the military) reliability (Evanzia, 1965) hindered introduction of frequency synthesizers for mobile communications. Aircraft and manpack terminals, particularly, require operation in severe temperature and vibration regimes as well as minimum operator skill and equipment maintenance.

The frequency range used for channelized communications by a host of mobile terminals reaches into the UHF frequency range (to approximately 900 MHz by a recent FCC decision). Hence, for a given channel width the desired frequency accuracy is greater than in the HF band.

Spurious emissions within adjacent channels for mobile applications are typically specified at least 80 dB below the carrier with a channeling time on the order of 1 sec. For collocated transceivers the spurious signal requirement increases to the order of 120 dB. Typical mobile synthesizers are described in Evers, 1966; Blachowicz, 1967; Hughes and Sacha, 1968; Avella and Perrigo, 1971; Beitman, 1971; Ribour, 1972.

3.   Radio Broadcasting

The large demand for radio channels has led to operating several transmitters on the same frequency. If the carriers are on the same frequency to within about 1 Hz, interference from multitone heterodyning is essentially eliminated (Stevens, 1954). Because such applications usually require that a transmitter be adjusted to the frequency of a received signal, the synthesizers require a resolution on the order of 1 Hz.

Many broadcast installations use several transmitters at once, each on a different carrier frequency. By successively switching a synthesizer to the different carrier frequencies, and by providing each transmitter with a frequency comparison loop whose time constant is much greater than that for the synthesizer to cycle through all frequencies in use, one synthesizer can control all of the transmitters (Van Duzer, 1964).

4. Television Broadcasting

Television transmitter carriers are ordinarily offset to reduce co-channel interference caused by reception of signals from more than one transmitter on the same channel, which occurs in the U.S. in areas equidistant from major population centers. This interference can be minimized by offsetting the carriers at multiples of the frame frequency with an accuracy on the order of 1 in 10^8 at VHF (Behrend, 1957; Middlekamp, 1958). In Europe, the irregular terrain in many countries permits operating many different television transmitters on the same channel, and all carrying the same program. Reception has been greatly improved by using carrier frequencies offset by 1/3, 2/3 or 4/3 of the line frequency. This improved performance was obtained with a relatively simple synthesizer designed for spurious signals below −80 dB relative to the carrier (Kroupa, 1967).

5. Atomic Frequency Standards

The quest for higher frequency stability has led to the development of frequency control systems whose natural frequency is determined by a molecular or atomic resonance, as opposed to the mechanical resonance of a quartz crystal (McCoubrey, 1966). The resonance frequencies for most such devices must be translated to a convenient output frequency, typically 5 MHz. In such applications, the required number of different output frequencies is fairly low (typically on the order of 100's), but spacing of the output frequencies is quite close. As an example, 1 in 10^{11} could be required, which is 5×10^{-5} Hz at 5 MHz. Because of the extremely narrow frequency tuning range of the output signal, a narrowband filter can follow the output of the device. Thus, the pressure for extremely pure spurious generation in the synthesizer is somewhat relieved. However, in general, these synthesizers must generate as little internal noise as possible so that the overall output frequency is determined primarily by the atomic reference. Furthermore, it has been shown that the shape of the power spectrum of the reference frequency limits the accuracy with which the frequency can be measured. Thus it is often desired that the frequency synthesizer not distort the frequency spectrum (Barnes and Mockler, 1960).

C. SYNTHESIZERS AS MEASUREMENT INSTRUMENTS

Although many measurements could be performed by a combination of a tunable oscillator and a frequency counter, measurement speed and accuracy, and hence, individual productivity can be increased greatly by introducing a frequency synthesizer. This permits using semiskilled measurement observers.

Frequency synthesizers operate in a relatively benign environment to make most measurements. Thus, it is often possible to achieve somewhat improved performance or lower cost at the expense of some limitation on the environmental range over which the instrument will work properly.

1.   Frequency Selective and Broadband Device Characterization

One of the more obvious applications for a frequency synthesizer is to measure the frequency transfer function of an active or passive device. For a simple magnitude measurement, synthesizer output power is measured on a wideband meter. The dynamic range and accuracy of such a measurement is limited by the synthesizer's spurious/noise emissions. For example, consider the measurement of more than 100-dB stopband attenuation in a filter with a nominal 100-kHz-wide passband. When the synthesizer is tuned into the stopband frequency region, noise and spurious signals transmitted through the 100-kHz passband of the filter must be considerably below −100 dB relative to the nominal synthesizer output level for an accurate measurement. Special attention to synthesizer design is necessary to meet such a performance requirement.

A frequency selective filter is usually inserted before the output measurement device to ease the difficulty of making such high dynamic range measurements. In the previous example, if such a predetection filter provided 50-dB rejection of frequencies in the passband of the filter under test, then the frequency synthesizer's spurious/noise emission performance could be relaxed by 50 dB. The equivalent of a tunable predetection filter can be obtained by mixing the output of the device under test against the synthesizer input test signal (Fig. 2). The mixing process translates the measured band to a very low frequency, permitting very narrow filter bandwidths to be achieved easily and with the feature that the narrowband filter is always centered on the frequency under measurement.

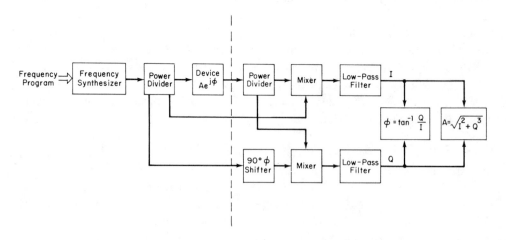

Fig. 2. Synchronous device measurement.

The useful switching speed for such a measurement is limited by the longest transient response in the system, which, in turn, corresponds to the narrowest bandwidth in the system. If the device to be measured has a 3-dB bandwidth of B Hz, then for a reliable measurement the synthesizer should dwell on the measurement frequency for several times $1/B$ sec. The actual required dwell time for a particular filter characteristic can be calculated from its known step response characteristics. In any event, there is little advantage for a synthesizer to have a switching time of less than about $1/B$.

2. Automated System Testing

In testing a complicated system, a large number of tests must be made to verify that it will work properly. Recent progress in automated testing is described in Schafer, 1973. In such applications, the switching speed of the synthesizer is important. Synthesizers not only permit automatic testing of the system, but also a large degree of self checking, which is important to establish confidence that the checkout is executed properly.

As a relatively sophisticated example, a computer-controlled, production test system for airborne phased array microwave modules (Dale and Howland, 1972) is described. The Reliable Advanced Solid-State Radar (RASSR) uses an antenna system composed of 1648 solid-state, microwave modules. Each identical module is in itself a small transmitter, receiver, and antenna capable of varying the phase of the RF signal. Because the antenna system must be capable of electronically (a) forming multiple beams, (b) steering the beam in azimuth and elevation, and (c) achieving low sidelobes, the RF modules must be matched in their phase and amplitude characteristics within precise limits over a broad range of operating conditions. Measurement of output power and phase as a function of input power, frequency, and temperature were required on a pulse basis. The measurements were further complicated by a frequency translation from S- to X-band. Other measurements included VSWR, output pulse characteristics, DC levels, receiver gain and phase characteristics, noise figures, spurious signal levels, and spectral purity. Each module required 700 measurements plus 800 limit comparisons.

For this test system a special coherent frequency synthesizer was developed to provide the desired high signal purity. The module test production program required approximately 10 minutes to execute, half of which time was used by the operator to read a spectrum analyzer. Major system errors due to mismatch, tracking, crosstalk, and directivity were removed by self-calibration techniques, supported by the frequency accuracy and repeatability of the coherent synthesizer.

3. Bridge Measurements

Bridge measurements, occasionally referred to as interferometric measurements, are used in a large number of fields. The basic circuit is given in Fig. 3. First, the bridge is balanced. Then the device to be tested is introduced or modified, either directly or by changing an environmental parameter such as frequency or magnetic field strength. The resulting bridge imbalance can provide extremely high measurement sensitivity (Montgomery, 1947).

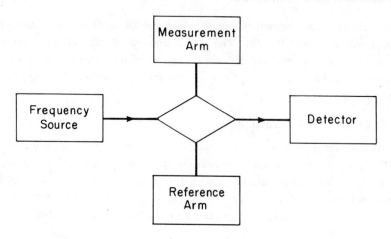

Fig. 3. Typical bridge measurements: small reflection coefficients, antenna patterns, radar cross sections, dielectric constants, loss factors, tensor permeabilities, microwave spectroscopy.

Typical applications of bridge measurements are of: (a) small reflection coefficients, (b) antenna patterns, (c) dielectric constants, loss factors and tensor permeabilities (e.g., Magid, 1968), (d) electron spin (paramagnetic) and nuclear magnetic resonance (microwave) spectroscopy, and (e) backscatter radar cross section. The last two areas provide interesting examples:

a. Microwave Spectroscopy

Microwave spectroscopy (Ingram, 1967) measurements may be divided into two classes. In the first class, the bridge is set up with the sample present, and the frequency source is varied to give a reflection or absorption spectrum. The other class, used especially for electron spin resonance (ESR) measurements, requires the additional imposition of a magnetic field on the sample. Traditionally, these latter measurements have been made with a fixed frequency while the magnetic field strength was measured (e.g., Wilmshurst, 1967; Alger, 1968). However, there are many ESR measurements where a constant magnetic field with varying frequency would be useful (Seurlock, Utton, and Wilmshurst, 1967).

The sample is usually placed in a cavity having a frequency dispersive characteristic. The dispersive characteristic of the bridge converts incidental FM on the reference signal into a corresponding amplitude modulation (noise), which limits the basic sensitivity of the bridge. This effect is minimized by centering the frequency of the cavity to the frequency of measurement, or conversely, by centering the frequency of measurement to the free-standing frequency of the cavity. However, it is clear that a necessary characteristic for high sensitivity measurements is that the frequency source have very low incidental FM (Buckmaster and Dering, 1967; Barnes, 1970).

b. Radar Cross-Section Measurements

A CW-balanced bridge is well suited to perform sensitive measurements of small targets at short distances, particularly in an indoor microwave chamber (Blore, Robillard and Primich, 1964). In this application, the sample arm of the bridge feeds an antenna that illuminates the target. A high degree of frequency stability is necessary to permit accurate and repeated balancing of the background reflections observed when the sample is not present. The radar cross section is then inferred from the change in reflected power due to bridge imbalance when the sample is placed in the environment. An example of an operational measurement range using this technique is given in Crispin and Siegel, 1968.

The accuracy of radar cross-section measurements using a CW-balanced bridge is limited by multiple-scattering paths. For example, incident energy from the radar can be reflected by the target to the chamber walls and the resulting backscattered energy can be reflected back to the radar receiver (Fig. 4). This source of degradation can be eliminated by frequency sweeping the radar reference signal. Then the resultant beat frequency out of the bridge is higher for reflections occurring at greater distances from the radar. Thus by suitable selection of the sweep rate and the IF selectivity characteristic of the radar receiver, the energy due to multiple scattering will not effect the measurement (Matsuo, 1963). The IF bandwidth should be narrow for highest sensitivity. Thus a frequency synthesizer is virtually mandatory to provide suitably high frequency stability in conjunction with a tightly controlled frequency versus time characteristic.

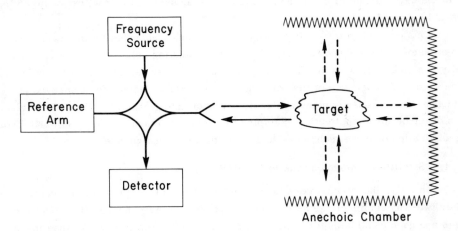

Fig. 4. Radar cross-section measurements.

4. Frequency Measurements

A traditional technique for measuring frequency is to compare the local standard frequency with the signal received from a standard frequency transmitter such as WWV. This is often accomplished by mixing the received radio frequency down to a very low frequency, and then counting the average rate of zero crossings in the result-ant beat frequency. In such applications, the frequency synthesizer provides an ad-justable fixed frequency translation ratio, and the basic long-term stability of the com-parison is determined primarily by the local frequency reference used to drive the frequency synthesizer (Finden, 1943; Van Duzer, 1964; Noyes, 1967).

5.   Frequency and Phase Tracker Testing

To verify proper operation of phase (or frequency) tracking systems, the experi-menter may generate a signal with controlled phase and frequency variations as a func-tion of time. Then the coefficients of the equivalent transfer function of the tracking system can be determined from the transient and/or steady state tracking errors. For example, the closed-loop damping coefficient and natural frequency ω_n of a second-order, phase-locked loop can be determined from the time-varying phase error ob-served after the input signal undergoes a frequency step. Typical phase-locked loop measurements are given in Gardner, 1966.

D.   ADVANCED SYSTEM APPLICATIONS

In most of the early utilization of the electromagnetic (and other) spectrum(s), the propagation media transmitted the signals with very little distortion. Thus the re-ceived signal was essentially a delayed, attenuated replica of the transmitted signal. However, the increased saturation of the radio frequency spectrum (e.g., Dean, 1970) and the increased requirements for reliable operation under adverse conditions, has led to the development of communications and radar systems that will operate on less favorable channels.

One major method for exploiting the less favorable channels uses successive transmissions on different center frequencies. Signaling schemes of this sort are really practical only with the use of synthesizers. These techniques will become much more widely used because of the continuously increasing demand for improved per-formance, and because synthesizer costs are decreasing rapidly. The following sec-tions describe three application examples in the electromagnetic spectrum and a poten-tial application of these techniques to an underwater sound channel.

1.   Satellite Communications to Mobile Terminals

Potentially, there are a very large number of mobile platforms that require higher two-way communications reliability than is available through HF. These plat-forms are good candidates for a satellite communications system. Because there could be thousands, even tens of thousands of such users, the terminal cost can easily dominate the total system cost.

For a large number of aircraft, the expense of mounting and pointing directional antennas implies that simple, inexpensive antennas with approximately hemispherical coverage are mandatory for practical total satellite system costs. The major disad-vantage of such an approach is that interfering signals are not discriminated against by antenna directivity. Included among such interfering signals are ground-reflected signals producing multipath (Fig. 5). In a simplified model for such a multipath channel (Fig. 6) the direct path is modeled as a delay of T_1 sec, and the path reflected via the ground is modeled as a delay of $T_1 + \Delta T$ sec with a relative amplitude of $\alpha \leqslant 1$. Thus, if s(t) is the transmitted signal, the receiver input signal is

$$\mu \left[s(t - T_1) + \alpha s(t - T_1 - \Delta T) \right] \quad ,$$

where μ is the scale factor due to free-space attenuation and antenna gain.

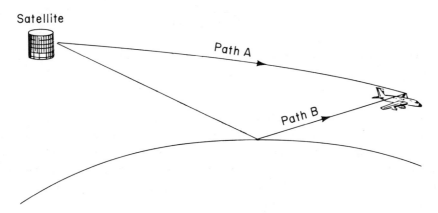

Fig. 5. Multipath channel.

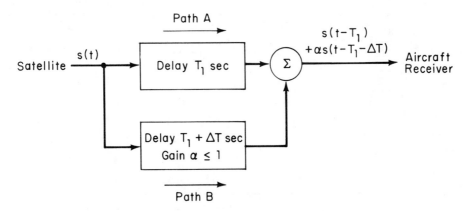

Fig. 6. Equivalent channel model.

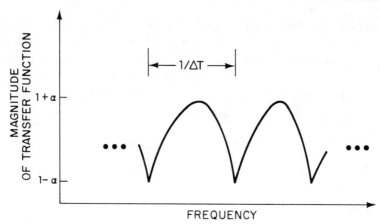

Fig. 7. Channel model transfer function; $1/\Delta T \simeq 360$ kHz for 5-deg satellite elevation and 30,000-ft aircraft altitude.

In the frequency domain the channel model has a corresponding transfer function (Fig. 7). The peaks occur at frequencies for which the reflected and direct path signals add; the nulls occur where the phase difference between the direct and reflected paths is 180 deg. To communicate reliably over such a channel, the signal must be spread out in the frequency domain so that adequate information is transmitted through the peaks of the channel transfer function.

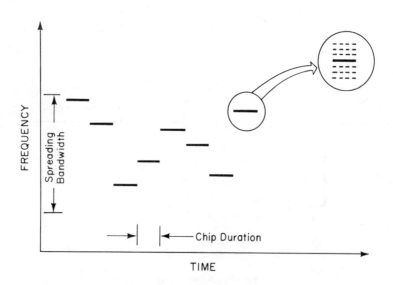

Fig. 8. Frequency-hopping signal structure.

An effective technique for communicating through such channels is to take succes-sive blocks of K information bits and to use each block to specify one of 2^K different frequency offsets from a nominal carrier frequency (Fig. 8). The transmitter changes the carrier frequency N times within the interval corresponding to the K information bits. Each of the N transmissions is called a "chip." Comparing with Fig. 7, if the carrier frequency is hopped over a bandwidth greater than the null-to-null frequency spacing of the channel transfer function, then some of the chips in each block will very likely pass through the channel.

The receiver hops in synchronism with the transmitter, thus effectively removing the hopping of the carrier frequency. It feeds the dehopped signal to filters matched to each chip, adds up the received energy corresponding to each of the 2^K possible se-quences of N chips, and decides that the transmitted signal (i.e., the K-bit block) was the one corresponding to the greatest energy sum (Fig. 9).

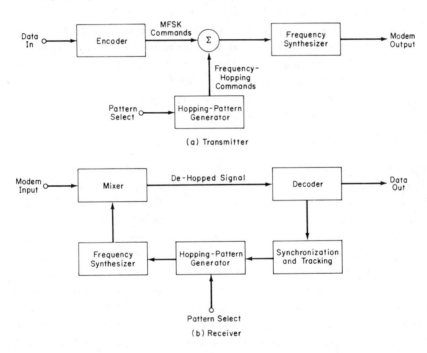

(a) Transmitter

(b) Receiver

Fig. 9. Frequency-hopping communication system.

Frequency hopping is also advantageous in rejecting other forms of interference, such as RFI, signals from other satellite users, and intermodulation noise. Such other techniques as pseudonoise (e.g., Judge, 1970) also enjoy many of these interfer-ence rejection properties. However, frequency hopping is less susceptible to narrow-band interference, and is by far the easiest to synchronize in the presence of signifi-cant initial timing and/or frequency errors.

Typical modulation schemes use approximately 1 chip per bit of information to be transmitted. Since the channel is effectively unavailable while the synthesizer is transmitting, it follows that the synthesizer switching time should be a small part of a chip duration. Thus for 2400-bps operation, synthesizer switching of 40 μsec or less is desirable. The resolution requirement for a receiving synthesizer can be inferred by observing that the bandwidth of each filter matched to a chip is essentially the reciprocal of the time duration of 1 chip. Thus the synthesizer should have a resolution at least an order of magnitude finer than the reciprocal of a chip duration, to limit degradation due to Doppler quantization error. For example, operation at 75 bps (TTY) with 1 chip per bit implies a synthesizer resolution on the order of 7 Hz is desirable.

The Tactical Transmission System (TATS) is an example of a system designed for communications in the presence of multipath and other interference by using frequency hopping (Drouilhet and Bernstein, 1969). This modem operates at data rates of 75 and 2400 bps with chip durations of 19 msec and 312 μsec, respectively. Successive chips can be hopped within a bandwidth of either 500 kHz or 10 MHz. The frequency synthesizers produce 2^{20} frequencies in response to a 20-bit digital command. The synthesizers switch from one frequency to another within 7 μsec, and yield less than 0.3 dB degradation when operating at 2400 bps. The spurious emission is better than 60 dB below the carrier within the hopping bandwidth.

2. Communications over Fading, Dispersive, Electromagnetic Channels

The previous example was based on a very simple channel model. For many channels there are many different paths over which the signal can travel, thereby introducing a very complicated form of multipath. Furthermore, the scattering centers may be moving so that the received signal has experienced a dispersion in frequency. Such channels occur, for example, when communicating by reflections from the moon (Green, 1968), by reflections from orbiting dipoles (Lebow, et al., 1964) and by forward scatter in the ionosphere or troposphere (Green, 1963).

These channels are generally characterized by two parameters (Kennedy, 1969). The first parameter, L, is the multipath spread. This is the duration of a received pulse when an impulse is transmitted (Fig. 10). The second parameter, B, called the Doppler spread, is the effective bandwidth of a received signal when a very narrowband

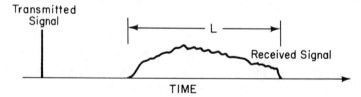

Fig. 10. Time dispersion (multipath spread).

sinusoid is transmitted (Fig. 11). In this latter case, the envelope of the received signal fluctuates with approximate duration $1/B$. The fading of two such signals becomes
independent as their frequency separation exceeds $1/L$.

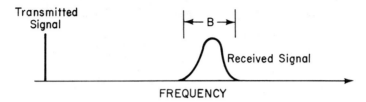

Fig. 11. Frequency dispersion (Doppler spread).

A method for communication over such channels is to transmit sequences of chips
of T-sec duration. The chip is a single sinusoid. The particular sequence of chips
corresponding to each message is specified by the particular modulation and coding
system. It can be shown that (a) if $L \leqslant 1/B$, best performance is obtained for
$L \leqslant T \leqslant 1/B$, and (b) if $L \geqslant 1/B$, best performance is obtained for $1/B \leqslant T \leqslant L$ (Kennedy
and Lebow, 1968).

Chips corresponding to the different possible messages are match filtered and envelope detected at the receiver. Detector outputs, fed to a decoder, decide what the
transmitted message was. The frequencies should be hopped over a bandwidth exceed-
ing $1/L$ so that the communications system, in effect, uses several independent channels. This is analogous to the bandwidth condition for the multipath channel discussed
in the previous section.

From these considerations a frequency synthesizer for communication on fading,
dispersive channels should at least be able to switch in a time that is small compared
to the larger of L or $1/B$. To provide the greatest possible number of independent
channels, the hopping bandwidth should be as large as possible. For Doppler tracking,
synthesizer resolution should be small compared to chip bandwidth. Note that as the
synthesizer switching speed requirement becomes more stringent with decreasing chip
duration, the resolution requirement becomes less stringent due to the corresponding
increase in chip bandwidth. Thus the required product:

(switching time) × (resolution)

is constant.

The Lincoln Experimental Terminal (LET), a complete, self-contained, air-
transportable ground terminal to test and demonstrate evolving space communication
techniques in a realistic environment, and provide efficient, multiplexed, digital communication on both coherent and time-varying dispersive channels with the moon and
active satellites, was an early application of the foregoing principles (Rosen and Wood,
1965; Drouilhet, 1965).

The elementary channel symbol, specified by a convolutional encoder, was a
200-μsec sinusoidal pulse transmitted on one of 16 offset frequencies (Fig. 12). The
received pulse was detected in 16 matched filters followed by envelope detectors. An

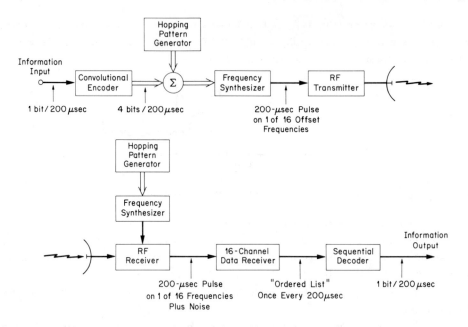

Fig. 12. Lincoln Experimental Terminal signal processing.

ordered list of the matched-filter outputs was passed to a sequential decoder. To prevent intersymbol interference on multipath channels, and to provide protection against other kinds of interference, the center of the 16 frequencies was changed from one pulse interval to the next in a pseudorandom fashion within a total bandwidth of 20 MHz.

The essential characteristics of the LET synthesizers were based on a requirement to cover 20 MHz with a resolution of 625 Hz. This resolution was necessary to permit sufficiently accurate tracking of the received frequency (a resolution of 1/8 of a matched-filter bandwidth). The center frequency of the synthesizer output band depended upon the particular synthesizer utilization within the terminal. A worst-case switching time specification of 10 μsec was imposed, assuming that 10-μsec dead time on each frequency change would result in a threshold degradation of about 0.5 dB. The achieved worst-case switching time was on the order of 5 μsec. A spurious output level of −40 dB, relative to the carrier, was considered adequate in this application.

3. Frequency-Agile Radars

Combining transmissions on several frequencies to obtain information is known as frequency agility when applied to radar. In addition to the rejection of such interference as multipath return, described in the previous two applications, frequency agility can reduce or eliminate the effects of target cross-section fluctuations as a function of frequency and/or time. Intuitively, this advantage is illustrated by observing that a target with two discrete scattering centers generates a multipath-channel characteristic similar to that described under Satellite Communications (Sect. D-1) (see Fig. 13). Target models are discussed in Skolnik, 1970.

TWO POINT RADAR TARGET

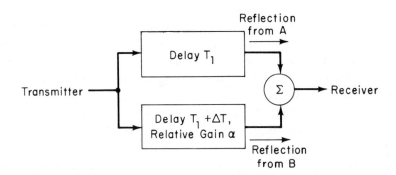

Fig. 13. Simplified radar channel model.

Frequency agility can be applied in two different ways. In the crudest approach, the radar makes successive measurements on different frequencies without combining the different returns. A more efficient, but more complicated method is for the radar to combine the returns from each of several different frequencies in roughly the same way as for communication systems. In addition to improved performance in a detection situation, studies (Ray, 1966) indicate that frequency agility will improve:

a. Degradation in tracking radars for homing or seeking, where the apparent location of the target varies at short range due to multiple scattering centers on the target.

b. Performance of conical-scan radars when the target scintillation has high frequency components.

c. Mapping radars where target recognition is of great significance.

Although there is a fair degree of similarity in the analysis and synthesis of frequency-agile radars and frequency-hopping communications, no available book describes the former class of systems in any detail. A bibliography of applicable papers is given in Barton, 1972.

The switching speed requirement for synthesizers in frequency-agile radars is determined by the allowable dead time between two successive transmissions. There is little advantage to switching faster than perhaps 1/10 of the minimum expected round-trip delay, because then the radar will be operational 90 percent of the available time.

The most severe short-term stability and spurious requirements for radars are probably imposed by coherent Doppler radars. This subject has been treated in great detail by Raven, 1966; Leeson and Johnson, 1966; Grauling and Healey, 1966. The basic requirement is that the noise level within the expected Doppler range (close to the carrier) be very low, so that small Doppler offset can be detected in the presence of strong background reflections.

4. Underwater Communications

The underwater channel provides an example of a somewhat dispersive multipath channel in which frequency hopping should provide substantially improved performance. However, such techniques do not appear to have been incorporated in actual hardware.

Time dispersion is caused by many signal paths from the transmitter to the receiver, due to multiple bottom and surface reflections as well as multiple refractions. For a path of 250 km, the time spread due to multipath can be on the order of 5 sec (Williams and Battestin, 1971). The Doppler spreading, usually quite small, is caused by relative motion of the transmitter, receiver, and reflecting interfaces (ocean surface). For carrier frequencies up to 5 kHz for fixed terminals the Doppler spread is less than 1 Hz even with a 20-knot surface wind and a shallow grazing angle. The differential Doppler experienced on different paths for moving systems can be as high as 30 Hz.

A possible signal design for moving systems would use chip durations within the range of 30 msec to 5 sec. The synthesizer for such an application is relatively simple since a switching time of several milliseconds is quite suitable. Even with 1-Hz resolution the total number of usable frequencies for long-range communications would only be on the order of 5000.

REFERENCES

R. S. Alger, Electron Paramagnetic Resonance: Techniques and Applications (Wiley, New York, 1968).

N. R. Avella and W. R. Perrigo, "AN/ARC-144 UHF Multimode Transceiver," Signal 26, 14 (1971).

J. A. Barnes, "Frequency Measurement Errors of Passive Resonators Caused by Frequency-Modulated Exciting Signals," IEEE Trans. Instrumentation and Measurement IM-19, 147 (1970).

J. A. Barnes and R. C. Mockler, "The Power Spectrum and Its Importance in Precise Frequency Measurements," IRE Trans. Instrumentation 9, 149 (1960).

J. A. Barnes, et al., "Characterization of Frequency Stability," IEEE Trans. Instrumentation and Measurement IM-20, 105 (1971).

D. Barrick, "Radar Measures Sea State 1,500 Miles Away," Microwaves, 12 (July 1973).

D. K. Barton, "Cumulative Index on Radar Systems," IEEE Trans. Aerospace and Electronic Systems AES-8, 1, 91-128 (January 1972).

W. L. Behrend, "Reduction of Co-channel Television Interference by Precise Frequency Control of Television Picture Carriers," IRE Trans. Broadcast Trans. Systems PGBTS-7, 6 (1957).

B. J. Beitman, Jr., "The AN/ARC-154(V) − A New Approach to Airborne Communications," NAECON 1971 Record, 47.

L. F. Blachowicz, "A Microelectronic Digital Synthesizer for UHF Communications," IEEE Conf. Publication 31, 91 (22-24 May 1967).

W. E. Blore, P. E. Robillard and R. I. Primich, "35 and 70 Gc Phase-Locked CW Balanced-Bridge Model Measurement Radars," Microwave J., 61 (September 1964).

H. A. Buckmaster, "The Fundamental Limit to the Balance of a Microwave Bridge Containing a Dispersive Element," IEEE Trans. Instrumentation and Measurement IM-16, 13 (1967).

R. L. Craiglow, "Frequency Control Techniques for Single Sideband," IRE Proc., 1697 (December 1956).

J. W. Crispin, Jr. and K. M. Siegel, Eds., Methods of Radar Cross-Section Analysis (Academic Press, New York, 1968).

L. S. Cutler and C. L. Searle, "Some Aspects of the Theory and Measurement of Frequency Fluctuations in Frequency Standards," IEEE Proc. 54, 136 (1966).

C. H. Dale and A. R. Howland, "Automated Test Equipment for Phased-Array Modules," IEEE Trans. Microwave Theory and Techniques MTT-20, 10 (1972).

W. Dean, Jr., "Electromagnetic Spectrum Control Programs," IEEE Trans. Aerospace and Electronic Systems AES-7, 23 (1971).

Electronics Buyer's Guide, 258 (McGraw-Hill, New York 1972).

Electronic Engineers Master, I, 168 (United Technical Publications, New York, 1973-4).

F. R. East, "The Frequency Synthesizer," Point-to-Point Telecommunication, 7, 6 (1962).

W. J. Evanzia, "Finally, the Armed Forces Get Solid-State Communications," Electronics, 63 (May 1965).

A. F. Evers, "A Versatile Digital Frequency Synthesizer for Use in Mobile Radio Communication Sets," Electronic Engineering, 296 (May 1966).

H. J. Finden, "The Frequency Synthesizer," J. IEE 90, Pt. 3, 165 (1943).

H. Flicker and J. Gerhold, "A Transistorized Frequency Synthesizer with High Spectral Purity for Short Wave Communication," IEEE Conf. Publication 31, 108 (22-24 May 1967).

F. M. Gardner, Phaselock Techniques (Wiley, New York, 1966).

P. E. Green, Jr., "Time-Varying Channels with Delay Spread," Radio Waves and Circuits, S. Silver, Ed. (Elsevier, New York, 1961).

P. E. Green, Jr., "Radar Measurements of Target Scattering Properties," Radar Astronomy, J. V. Evans and T. Hagfors, Eds. (McGraw-Hill, New York, 1968).

C. H. Horn, "A Carrier Suppression Technique for Measuring S/N and Carrier/Sideband Ratios Greater than 120 dB," Proc. 23rd Ann. Symp. Frequency Control, U.S. Army Electronics Command, Fort Monmouth, N.J., 223 (6-8 May 1969).

R. J. Hughes and R. J. Sacha, "A Miniature Precision Digital Frequency Synthesizer," Proc. 23rd Ann. Symp. Frequency Control, U.S. Army Electronics Command, Fort Monmouth, N.J., 211 (6-8 May 1969).

R. J. Hughes and R. J. Sacha, "The LOHAP Frequency Synthesizer," Frequency, 12 (August 1968).

D. J. E. Ingram, Spectroscopy at Radio and Microwave Frequencies (Plenum Press, New York, 1967).

W. J. Judge, "Tactical Radio Communications Privacy – One Approach," 1970 Carnahan Conf. Electronic Crime Countermeasures, Lexington, Ky., 231.

R. S. Kennedy, Fading Dispersive Communication Channels (Wiley, New York, 1969).

R. S. Kennedy and I. L. Lebow, "Signal Design for Dispersive Channels," IEEE Spectrum, 231 (March 1964).

V. Kroupa, "Single-Frequency Synthesis and Frequency-Coherent Communication Systems," IEEE Conf. Publication 31, 96 (22-24 May 1967).

V. Kroupa, "Theory of Frequency Synthesis," IEEE Trans. Instrumentation and Measurement IM-17, 56 (1968).

J. Larcher, T. R. T. France, and J. Noordanus, "Frequency Synthesizers for Radio Equipment," Philips Telecom. Rev. 27, 149 (1968).

O. P. Layden, "Frequency Control for Tactical Net SSB Equipment," Proc. 23rd Ann. Symp. Frequency Control, U.S. Army Electronics Command, Fort Monmouth, N.J., 14 (6-8 May 1969).

I. L. Lebow, K. L. Jordan, Jr., and R. P. Drouilhet, Jr., "Satellite Communications to Mobile Platforms," Proc. IEEE 59, 139 (1966).

D. B. Leeson, "Short-Term Stability for a Doppler Radar: Requirements, Measurements and Techniques," Proc. IEEE 54, 244 (1966).

S. Leinwoll, "The Problem of Congestion in High-Frequency Broadcast Bands," IEEE Trans. Broadcasting BC-14, 56 (1968).

M. Magid, "Precision Determination of the Dielectric Properties of Nonmagnetic High-Loss Microwave Materials," IEEE Trans. Instrumentation and Measurement IM-17, 291 (1968).

E. T. Martin and G. Jacobs, "The Future of Shortwave Broadcasting," IEEE Trans. Broadcasting BC-14, 95 (1968).

M. Matsuo, "Backscattering Measurements by an FM Radar Method," IEEE Trans. Antenna Propag. AP-13, 485 (July 1963).

A. O. McCoubrey, "A Survey of Atomic Frequency Standards," Proc. IEEE 54, 116 (1966).

L. C. Middlekamp, "Reduction of Co-channel Television Interference by Very Precise Offset Carrier Frequency," IRE Trans. Broadcast Trans. Systems PGBTS-12, 5 (1958).

D. L. Moffatt, "Impulse Response Waveforms of a Perfectly Conducting Right-Circular Cylinder," Proc. IEEE 57, 816 (1969).

C. G. Montgomery, Ed., Technique of Microwave Measurements (McGraw-Hill, New York, 1947).

M. E. Peterson, "The Design and Performance of an Ultra-pure, VHF Frequency Synthesizer for Use in HF Receivers," Proc. 25th Ann. Symp. Frequency Control, U.S. Army Electronics Command, Fort Monmouth, N.J., 231 (26-28 April 1971).

M. E. Peterson, private communication, 1972.

R. S. Raven, "Requirements on Master Oscillators for Coherent Radar," Proc. IEEE 54, 237 (1966).

J. Ribour, "Noise Characteristics in Synthesizers for Mobile Transmitter/ Receivers," Electrical Communication 47, 2, 117-126 (1972).

A. St. J. Reynolds, "A Method of Assessing the Short-Term Instability of Frequency Synthesizers," IERE Radio Receivers and Associated Systems (4-6 July 1972).

V. M. Rozov, "Precision Frequency Equipment," Elektrosvyaz 7, 33 (1959).

S. Runyon, "Focus on Signal Generators and Synthesizers," Electronics Design 10, 62-73 (May 10, 1973).

G. E. Schafer, "U.S. Automated Test Instrumentation Progress," Microwave J., 27-29 (April 1973).

R. B. Shields, "Review of the Specification and Measurement of Short-Term Stability," Microwave J., 49 (June 1969).

V. E. Van Duzer, "Notes on the Application of Frequency Synthesizers," Hewlett-Packard J. 15, 7 (1964).

B. Weitz, "Selection Checklist – Frequency Synthesizers," Electronic Products, 125 (February 15, 1971).

R. E. Williams and H. F. Battestin, "Coherent Recombination of Acoustic Multipath Signals Propagated in the Deep Ocean," J. Acoustical Soc. of Am. 50, 1433 (1971).

N. H. Young and V. L. Johnson, "Design Principles of High Stability Frequency Synthesizers for Communications," IRE WESCON Conf. Rel. 1, 35 (1957).

Chapter II
Contemporary Frequency Synthesis Techniques
B. H. Hutchinson, Jr.

This chapter seeks to provide an understanding of the three basic frequency
synthesis techniques of current usefulness and importance:

a.   Direct synthesis using iterated identical stages (sometimes called
     mix-filter-divide),

b.   Programmable divide-by-N and phase lock with variations,

c.   Digital table look-up of sinusoidal sample values.

How these three types of synthesizers work and the advantages, disadvantages and
design tradeoffs inherent in each are discussed in detail.   The material presented is
based on my confrontations with spaces in satellite communications system diagrams
labeled FREQUENCY SYNTHESIZER that were filled with new designs of all three
types of synthesizers.

No attempt is made to discuss or list all known frequency synthesis techniques
with design examples and references.   Many good survey papers are available as
well as detailed descriptions of particular designs.   My purpose is to fill the gap between these two by concentrating on the important features of what appear to be the
leading candidates for synthesizer design.

The contemporary relevance of the three techniques described is based primarily
on fundamental advances in solid-state component technology that have occurred in the
past 12 years or more.   The multistage iterative direct synthesizer, conceived in
1949,[1] was made practical around 1960 by a mature solid-state discrete component
technology.   The divide-by-N, phase-lock synthesizer became practical with the
advent of high-speed, small-scale, digital integrated circuits in the middle 1960s.[2,3]
The digital table look-up synthesizer was made practical in the late 1960s[4] by MSI/LSI
techniques applied to semiconductor read-only memories (ROMs).   Since all three
techniques involve digital programming or commanding, they can all by viewed as
growing out of the "digital revolution" that started in the 1950s and is still with us.

In this light, it is easy to see how earlier frequency synthesis techniques to
select a frequency (not discussed here) followed naturally from knob-turning to
vacuum-tube components to the world of analog systems.   Examples include the drift-
canceled oscillator harmonic selector, multiple harmonic-locked oscillators providing variable offset frequencies inside one large phase-locked loop, and direct synthesis without interstage dividers.[5,6]

Emphasized then, is (1) choosing a technique, and (2) designing to meet a particular set of requirements efficiently, rather than (3) choosing among many commercially available synthesizers.   The third course is often best, especially where
requirements may change widely, as in a highly versatile computerized automatic test

system, or where decimal organization and convenience are desired. Selection of commercial units is simplified by synthesizer manufacturers' catalogs and application notes, which are voluminous and generally excellent. However, it is usually fairly obvious when a custom design is simpler or cheaper than a commercial unit or will provide better performance.

A. DIRECT SYNTHESIS WITH ITERATED IDENTICAL STAGES

1. The Principle

The divider is the key in the basic principle of iterative direct synthesis (Fig. 1) to make possible identical stages that share a common set of K fixed continuous component frequencies that permit the output of each stage to interpolate precisely between the frequency steps of size Δf, selected by the RF switch of the next stage.

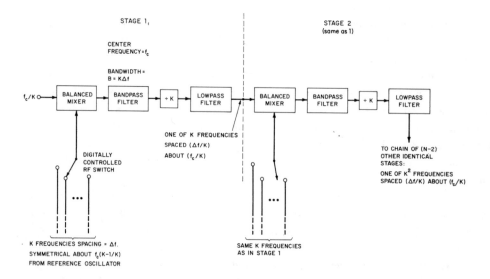

Fig. 1. Iterative direct synthesis: first two of N identical stages.

The component frequencies are derived from the input reference frequency by generating a comb or line spectrum with space Δf, and then using a bank of K crystal filters, or sometimes fixed phase-locked oscillators, to select individual component frequencies. The RF switch (indicated as a selector switch for clarity, Fig. 1) is usually built as K separate ON/OFF diode gates with control logic to turn on only one of the K gates under control of $\log_2 K$ bits (4 BCD bits, if K = 10). Sometimes a drift-canceled oscillator or tunable phase-locked oscillator is used to combine component-frequency filtering and switching in the stages of a direct synthesizer, an important exception to the notation that these techniques are now little used.[7] The bandpass filter merely selects the desired mixer output frequency, usually the sum frequency. The lowpass filter eliminates harmonics in the output of the divider, which in recent designs is nearly always an integrated-circuit counter. The output of the last stage of the basic synthesizing chain is taken directly from the last band-pass filter.

A most popular variation[8] to the basic arrangement uses two mixers and band-pass filters, allowing the RF gates to operate at a frequency much lower than f_c, where better ON/OFF ratios can be obtained. Basically, the plan achieves improved spurious suppression in exchange for increased complexity.

Typical values of f_c lie between a few megahertz and a few tens of megahertz, while Δf values usually range from tens to hundreds of kilohertz.

2. Parameter Tradeoffs

The first design decision is the division ratio – K (Fig. 1). For manual control applications such as test equipment, K is nearly always 10. Where the synthesizer is always controlled by a digital machine, as in frequency-hopping communications systems such as LET[9] or TATS,[10] a power of 2 makes more sense. In practice, K = 4 is advantageous. Compared to K = 8 it gives a simpler stage and fewer component frequencies to generate, plus larger percentage bandwidth and slightly faster switching time. The only disadvantages are that 1.5 times as many stages are required, and spurious in-band mixer outputs are of lower order; however, neither consideration is very troublesome. A K of 16 offers no real advantage in exchange for the complex RF switch matrix, component frequency generator and limited percentage bandwidth.

A value of K = 2 provides maximum percentage bandwidth and switching speed, but spurious second- and third-order mixer outputs will fall in-band, severely limiting spectral purity ($\sim$40 dB). Such a choice should be limited to highly specialized applications.

The basic characteristics of a synthesizer consisting of a chain of N stages (Fig. 1) are easily determined from the stage parameters:

$$\text{Number of output frequencies} = K^N$$

$$\text{Output bandwidth} \triangleq B = K(\Delta f)$$

$$\text{Minimum output frequency increment} = \frac{\Delta f}{K^{(N-1)}} = \frac{B}{K^N} \quad .$$

The exponential increase in the number of output frequencies with the number of stages makes it clear that sheer number of frequencies, and hence, fine output resolution with fixed B, come easy in a direct-chain synthesizer. These quantities are thus involved in basic design tradeoffs in a relatively minor way.

Fractional stage bandwidth B/f_c is harder to increase.

For a given filter center frequency, f_c and K value, maximum bandwidth, B is constrained by the need to pass the upper sideband output of the mixer while

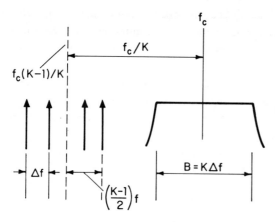

Fig. 2. **Filtering constraint for iterated direct synthesis.**

rejecting the mixer input frequency from the RF switch array, assuming the desired spurious output suppression exceeds the achievable guaranteed mixer balance, as is nearly always the case (see Fig. 2, drawn for K = 4). This filtering constraint (Fig. 2) can be stated as follows (assuming for the moment an ideal rectangular filter passband):

$$(\frac{K-1}{2}) \, \Delta f + \frac{B}{2} \leqslant \frac{f_c}{K} \quad .$$

Substituting $\Delta f = B/K$,

$$B \leqslant f_c \, (\frac{2}{2K-1}) \quad .$$

Since percentage bandwidth = $\dfrac{100 \, B}{f_c}$,

$$\text{Percentage bandwidth} \leqslant \frac{200}{2K-1} \text{ (ideal)}$$

$$\leqslant 28.6\% \text{ for K = 4}$$

$$\leqslant 13.3\% \text{ for K = 8}$$

$$\leqslant 10.5\% \text{ for K = 10} \quad .$$

Actual filter passbands are, of course, nonrectangular; furthermore, the closest achievable approximation to a rectangular passband would be most undesirable, because the increased envelope delay of such a filter would ruin frequency switching time – complexity aside. Hence, a 2:1 stopband/passband ratio is a reasonable lower bound, which gives:

$$\text{Percentage bandwidth} \leqslant \frac{100}{2K-1} \text{ (practical)}$$

$$\leqslant 14.3\% \text{ for K = 4}$$

$$\leqslant 6.7\% \text{ for K = 8}$$

$$\leqslant 5.25\% \text{ for K = 10} \quad .$$

The frequency switching time of one stage is simply the delay between an abrupt change in the digital command to the RF switch and the completion of the resulting frequency change at the lowpass filter output – call it T_S. Then

$$T_S = T_{(BPF)} + T_{\div} + T_{(LPF)} \quad .$$

Overall N-stage synthesizer switching time $= N\, T_S$ (to within the minimum increment, B/K^N). The delay of the RF switch is easily made small relative to the other delays; furthermore, it adds to $N\, T_S$, not to T_S, and so is not considered further. $T_{(BPF)}$ and $T_{(LPF)}$ are the worst-case envelope delays of the two filters:

$$T_{(BPF)} = C_1/B$$

$$T_{(LPF)} = C_2/f_{co} \quad ,$$

where $f_{co} = [f_c + (B/2)]/K$, the cutoff frequency. The divider delay, $T_{\div}$, is essentially a half-cycle of f_{co} for a counter with square-wave output; since it is small and varies as $1/f_{co}$, it is henceforth considered as absorbed into C_2. C_1 and C_2 are envelope-delay constants of proportionality. Given a choice of filter stopband attenuation and stopband/passband ratio (e.g., Fig. 2), the filter types and number of poles are determined, and hence, so are C_1 and C_2; they are not functions of B or f_c. Typically, $C_1 \approx 1$ and $C_2 \approx 0.5$ for frequencies in megahertz and envelope delay in μsec, representing filters with about five poles each.

Substituting for f_{co}, and also substituting:

$$B = B_{max} = \frac{f_c}{2K - 1}$$

(the practical value based on 2:1 stopband/passband ratio):

$$T_{(BPF)} = \frac{C_1 (2K - 1)}{f_c}$$

$$T_{(LPF)} = \frac{C_2 K}{f_c} \left[\frac{2 (2K - 1)}{1 + 2 (2K - 1)} \right] \approx \frac{C_2 K}{f_c} \quad \text{with a few percent for } K \geqslant 4,$$

hence

$$T_S \approx \frac{C_1 (2K - 1) + C_2 K}{f_c} = \frac{2K - 1 + \frac{1}{2} K}{f_c} = \frac{2.5K - 1}{f_c}$$

for $C_1 = 1$, $C_2 = 0.5$.

Having identified a major tradeoff, overall synthesizer switching time, $N\, T_S$, is reduced by increasing f_c, the stage center frequency at which mixer, bandpass filter, divider, and (approximately) RF switch must operate. Achievable output bandwidth, B, also increased with f_c, as derived. It is also apparent that T_S is reduced directly by a lower value of K for a fixed f_c. However, if the number of output frequencies is to remain the same, N must increase as K decreases; hence, N T_S is proportional to $K/\log_2 K$ rather than K. For example, N T_S is reduced by a factor of 1.33 if K is reduced from 8 to 4.

The spurious output suppression achievable in a direct synthesizer is not predictable, even approximately, by an analysis such as that used for switching time: it is very much a function of the care and effort spent on circuit design and packaging. Obviously, the suppression of unwanted output sidebands can never exceed the ON/OFF ratio of the RF switches or the sideband suppression at the component frequency outputs. It often happens, however, that the spurious output suppression is significantly worse than the ON/OFF ratio for a single RF switch circuit or an isolated component frequency output, because of stray coupling among the many RF switch circuits and component frequency distribution lines. Ordinary, routine, first-cut printed board packaging will rarely provide better than 40 to 50 dB spurious suppression for an f_c measured in megahertz to tens of megahertz. Reasonable care in layout can provide 60 to 70 dB, while very careful design and fabrication techniques can and have achieved 80 to 100 dB spurious output suppression.[8,11]

Good mixer balance and stopband rejection in the filters is also important. Even if a mixer output signal does not fail in the desired output band, it can produce a spurious in-band signal after going through the divider. This is because a frequency divider reduces the level of sidebands at its input by 20 $\log_{10} K$ dB, but does not change the sideband offset frequency. For example, if a divide-by-4 circuit is presented with an input signal at 8 MHz and a sideband 4.1 MHz away at 3.9 MHz, which is 40 dB down, the main output is at 8/4 = 2 MHz. The sideband will still be 4.1 MHz lower, at 2 − 4.1 = −2.1 MHz, or just 2.1 MHz if unconcerned with phase. Thus, the divider output has a sideband 100 kHz away, suppressed by 40 + 20 $\log_{10}(4)$ = 52 dB.

Though hard to quantify, it is obviously true that the good circuit performance and low stray coupling required to minimize spurious output levels is harder to achieve at higher values of f_c. Thus there is a tradeoff between (a) higher f_c to reduce switching time and increase bandwidth, and (b) lower f_c to make spurious suppression easier.

3. Advantages of Direct Iterative Synthesis

The unmatched versatility and flexibility of this basic synthesis scheme (Fig. 1) for very demanding applications is one reason for its popularity in general-purpose test instruments. If effort is put into careful design and construction, the iterative direct synthesizer can deliver, if desired, good performance in <u>all</u> areas simultaneously:

a.	Fast frequency switching	Microseconds or less
b.	Almost arbitrarily fine resolution	10^{-2}-Hz steps; finer if desired
c.	Good to excellent spurious suppression	60 to 100 dB

    d.   Phase noise   Close to reference-oscillator limited

    e.   Wide bandwidth   Hundreds of megahertz.

Some commercial general-purpose synthesizers approach such a combination (price commensurate with performance, of course).[12] Usually, high performance is demanded in one or two areas while reasonable performance is retained in other areas. For example, the design described in Chapter III by Gundel combines extremely fast switching (1.5 μsec) with wide output bandwidth (tens of MHz) while retaining reasonably fine resolution and quite respectable spurious suppression (70 dB).

The iterative direct synthesizer is inherently modular, and its frequency programming separates naturally into distinct decades (octades or quatrades). This makes it easy to assemble units with different capabilities from a few basic modules, and to transform the units as requirements change. Also, it is easy to substitute a voltage or mechanically tuned oscillator for any stage, obtaining a tunable oscillator with arbitrarily programmable center frequency and bandwidth.

    4.   Disadvantages

The excellent all-around performance and versatility of the direct iterative synthesizer is bought at a price: a high number of digital and RF circuitry parts. Monolithic integrated circuits cannot provide the necessary filters, mixers and complete RF switching networks in amplifiers and dividers. Furthermore, thick-film hybrid circuitry cannot always deliver the required excellent RF isolation and low impedance interconnections and is not at its best with the wide variety of necessary components (inductors, transformers). Microwave-precise integrated circuits on low-loss ceramic dielectric substrates have been applied successfully in UHF portions of some commercial direct synthesizers; however, the basic complexity disadvantage remains.

Also, RF packaging and interconnection can be bothersome if stray coupling is controlled well enough to achieve even reasonable spurious suppression (anything better than 50 dB). This becomes especially difficult if better than 70-dB suppression is wanted as indicated by the special techniques in successful designs: RF switch components encased in conformal metal castings, multiple-layer metal shields and compartments, extensive power decoupling. Also, spurious outputs can occur anywhere across the output band.

The iterative direct synthesizer is not well suited to large percentage bandwidths. Hence, effort is required to obtain a large absolute output frequency range. A high center frequency and correspondingly greater bandwidth for all stages makes the circuit design more difficult and leads to greater power consumption, cost, and stray coupling problems. Frequency multiplication of the basic chain output is often used, but the spurious signal and noise suppression is degraded by $20 \log_{10} M$, where M is the multiplication, and hence, bandwidth-expansion factor. A special wideband, high-frequency output section is also often used, sometimes including downconversion, to obtain a large percentage bandwidth or frequency coverage starting at or near zero.

Potentially, this is the best solution for performance reasons; however, complexity, power consumption, and noise increase accompanied by a departure from an optimum arrangement of identical cascaded modules.

B. PROGRAMMABLE DIVIDE-BY-N PHASE-LOCK SYNTHESIS

1. The Principle

Considering the simplest divide, phase-lock synthesizer (Fig. 3), where $P = 1$ and f_d is not used, the loop comes to equilibrium at a fixed phase error only when the divider output frequency is precisely equal to f_{ref}, since frequency is the time derivative of phase. Therefore, the divider input frequency, and hence, the VCO output

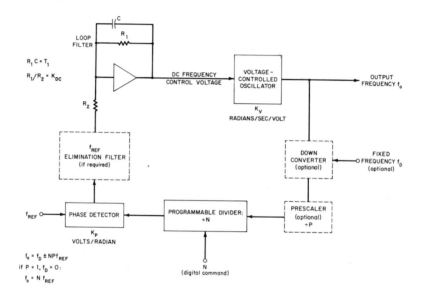

Fig. 3. Programmable divide-by-N phase-lock synthesizer.

frequency must be precisely $N f_{ref}$, and the equilibrium phase error voltage is exactly that required to tune the VCO to that frequency. The minimum frequency increment is equal to f_{ref}, produced when N is changed by 1.

Since the divide-by-N counter was made practical by high-speed integrated circuits, the programmable divider can be built in many ways using various forced reset schemes and/or comparators. The technique that affects speed the least is to selectively inhibit resetting of some bits following the all-1's state, N_{max}, under control of the input command, so that $(N_{max} - N)$ follows N_{max} in the count cycle.

Since the worst reset delay (one flip-flop delay plus one or two gate delays) for all bits in the divide-by-N counter must be less than a half cycle of the counter input frequency, a high-speed prescaler of a few bits and/or a downconverter are often used to reduce the divider speed requirement. The prescaler reduces the required f_{ref}, and hence, the loop bandwidth achievable by the factor P, if the minimum frequency increment is kept the same. The downconverter adds complexity and limits the achievable percentage bandwidth compared to direct count, but can reduce close-in spurious/noise sidebands, and permits output frequencies in the lower microwave region with an ordinary integrated-circuit, divide-by-N.

2.   Design Tradeoffs

The closed-loop, response function of the phase loop, i.e., ϕ out/ϕ ref (Fig. 3)

$$F(s) = \frac{1}{T_1 s + 1} \quad \text{(accounting separately for } K_{DC}) \quad .$$

Substituting,

$$\frac{\phi \text{ out}}{\phi \text{ ref}} = \frac{N \, K_L}{T_1 s^2 + S + K_L \exp[-s\tau]} \quad .$$

The overall loop is thus seen to have a second-order, response function. Loop parameter selection is now simply a matter of ensuring that this closed-loop, response function has left-half-plane poles and adequate damping. A trick simplifies the process greatly. Let $s = \alpha + j\omega$ at the poles of the loop-transfer function; assume that $\alpha < 0$ for stability, and that $|\alpha| = |\omega|$ for simplicity. Substituting in the expression for (ϕ out/ϕ ref) and setting real and imaginary parts of the denominator equal to zero separately, as must be true at the poles, these expressions are obtained (Fig. 3):

$$K_L = \frac{\omega}{\exp[\omega\tau] \cos \omega\tau}$$

$$T_1 = \frac{1 - \tan \omega\tau}{2\omega}$$

where

ω = radian frequency of the closed-loop poles

T_1 = loop filter RC time constant

$\tau = \dfrac{1}{f_{ref}}$ = divider delay

K_L = loop gain = $\dfrac{K_P \, K_V \, K_{DC}}{N}$ (N is replaced by NP if a prescaler is used)

The assumption of equal real and imaginary parts at the poles, resulting in simple equations, also produces a closed-loop, response function resembling the second-order Butterworth; a practical sort of filter characteristic with a well-behaved transient response that makes a good design value, but the transient response is increasingly underdamped.

The key design choice is ω, given f_{ref}; the 3-dB, closed-loop, bandwidth B is approximately $(\omega/2\pi)\sqrt{2}$ Hz. Given ω and the foregoing equations, K_L and T_1 are determined easily. The quantity $\exp[\omega\tau]\cos\omega\tau$ is within 10 percent of $\sqrt{2}$ for the usual range of loop bandwidths, 5 to 15 percent of f_{ref}. Thus, $K_L \approx \pi B$. The quantity K_L can be parceled out among K_p, K_v, K_{DC} in any convenient way, and the basic synthesizer loop design is finished. Note that K_L depends on N; often a D/A converter is connected to the command input to program K_{DC} or K_p to keep K_L constant at the design value as N varies, and also to allow for VCO nonlinearly (varying K_v).

A D/A converter can also provide an open-loop, pretune voltage to the VCO to allow arbitrary output bandwidth while keeping the steady-state, phase error small. Without pretuning, the maximum tuning range of the phase-lock synthesizer corresponds to a steady phase error of $\pm(\pi/2)$, assuming the usual sampled-sawtooth or duty-cycle phase detector with a linear characteristic. At the extremes of the locking range, the control voltage supplied to the VCO is $\pm(\pi/2) K_p K_{DC}$, so that its tuning range is

$$\pm(\pi/2) K_p K_{DC} K_v = \pm N K_L (\pi/2) \text{ radians/sec}$$

$$= \pm \frac{N K_L}{4} \text{ Hz} \quad .$$

Since $K_L \approx \pi B$,

$$\text{Loop hold-in range} \approx \pm(\pi/4) N B \text{ Hz}$$

where B = closed-loop, 3-dB bandwidth in Hz, an intuitively satisfying and plausible result.

Obviously, the maximum achievable bandwidth with the Butterworth-like response corresponds to $\omega\tau = (\pi/4)$ and $T_1 = 0$, at which point $B \approx (\omega/2\pi)\sqrt{2} = 0.177 f_{ref}$. The corresponding maximum K_L is $0.506 f_{ref}$. Higher K_L, with T_1 still zero, results in wider bandwidth loop response, but with a peak just before the cutoff frequency and overshoot in the transient response; hence, no large improvement in bandwidth can be obtained this way without endangering loop stability.

All the typical loop parameter values for f_{ref} = 100 kHz, scale linearly with f_{ref} (T_1 inversely, of course) (Table 1). With a closed-loop bandwidth well under 5 percent of f_{ref}, the effect of the divide-by-N delay on loop parameters is small.

The main effects of the loop feedback on noise (external pickup, internal circuit or VCO noise) is illustrated by a simple model. Assume an unwanted sinusoidal interfering voltage, V_n (peak), added at the phase detector output with K_{DC} = 1, for convenience. Then if the feed-back noise voltage out of the phase detector is V_p, then V_v, the noise at the VCO input, is the vector sum of V_p and V_n. Further, if the frequency of the interference, $\omega_n/2\pi$, is small compared to the loop bandwidth, V_v and V_p are in phase quadrature because of the K_v/s phase transfer function of the VCO.

Table 1. f_{ref} = 100-kHz Loop Parameters

Approximate Closed-Loop, 3-dB Bandwidth (kHz)	Loop Gain, K_L ($\times 10^4$)	Loop Filter Time Constant, T_1 (μsec)	Loop Filter Attenuation, f_{ref} = 100 kHz (dB)	Delay Factor $\exp[\omega\tau]\cos\omega\tau$
5.6	2.0	15.0	19	1.243
10.5	3.3	4.7	10	1.425
14.1	4.1	2.2	5	1.52
17.7	5.06	0	0	1.55

Note: Bandwidth, B, values can be read as percents of f_{ref}, and T_1 and K_L can be scaled linearly to any other value of f_{ref}.

This simple quadrature model gives results accurate to ~2 dB for frequencies up to half the closed-loop bandwidth. Hence:

$$|V_n|^2 = |V_v|^2 + |V_p|^2 = |V_v|^2 \left[1 + \frac{K_p K_v}{\omega_n N}\right]^2$$

$$|V_n| \approx \frac{|V_v| K_p K_v}{\omega_n N} = \frac{|V_v| K_L}{\omega_n} \quad .$$

Recall that $K_{DC} = 1$ was assumed for convenience.

The noise modulation index of the VCO is

$$\Delta = \frac{V_v K_v}{\omega_n}$$

and the level of each of the resulting two symmetrical spurious sidebands relative to the desired output is $20 \log_{10}(\Delta/2)$. Substituting,

$$\text{Spurious sideband level} = 20 \log_{10}\left(\frac{K_v V_n}{2K_L}\right)$$

(at any spot frequency half the loop bandwidth or less).

Thus, the spurious sidebands due to V_n are better suppressed as the quantity K_v/K_L becomes smaller. In other words, K_v, the VCO tuning sensitivity, should be as small as possible for a given total loop gain to minimize sidebands due to circuit noise or pickup, which is intuitively plausible. Note also that

$$\frac{K_v}{K_L} = \frac{N}{K_p} \text{ (with } K_{DC} = 1\text{)}$$

Hence, an equivalent statement is that small N and large K_p reduce sideband noise for a given K_L. Now K_p is hard to increase much beyond 6 radians/volt, which implies ±10 V output swing for ±(π/2) radians phase error. 1 to 3 rad/V would be more typical, and there is little reason for significantly smaller values. In practice, then, the only way to improve noise performance is to reduce N. If the output frequency and minimum frequency increment are to remain the same, a downconverter must be introduced (Fig. 3). This is most attractive when the percentage bandwidth is not too large; with an octave output bandwidth and one value of f_d, for example, N could be reduced by less than a factor of two, and spurious mixer outputs would not be a problem.

Since N is replaced by NP if a fixed prescaler is used, prescaling clearly increases the sideband level due to circuit noise or pickup, in addition to the reduction in f_{ref}, and hence, loop bandwidth and K_L if the minimum frequency increment $P f_{ref}$ is to remain constant for $P > 1$.

The quantity $K_v V_n$ in the spurious sideband expression has dimensions of radians/ sec, and is, in fact, the peak VCO frequency deviation caused by V_n if applied directly to the VCO input with no loop feedback. The internal FM noise of the VCO can, of course, be modeled by an equivalent input, V_n, from which it follows that VCO FM noise is reduced by a factor of $2K_L$ at frequencies less than half the closed-loop bandwidth. Since K_L is directly proportional to loop bandwidth, wide loop bandwidth clearly improves suppression of VCO FM noise, and also increases the maximum frequency at which suppression is effective.

The foregoing analysis is not affected much if K_{DC} is made greater than unity with K_L fixed to reduce K_v. Replace the VCO with an equivalent in which $K_v' = K_v K_{DC}$ and the sensitivity to voltage V_n is the same as before. The point of sensitivity has merely moved from VCO input to amplifier input; not necessarily an advantage.

For noise or interference frequencies greater than the loop bandwidth, the loop feedback has little or no effect. Hence, a V_n injected at some point in the DC control circuitry will merely produce a V_v attenuated by whatever lies between the injection point and the VCO input, if anything (e.g., the RC loop filter and/or the f_{ref} elimination filter). The resulting sideband level can be calculated from the foregoing expressions involving modulation index Δ.

A wide loop bandwidth also reduces the time required for the synthesizer to settle on a new frequency following a change in the input command, N. Exact analysis can be complicated, but in practice, a fairly simple rule of thumb has been found adequate:

$$\text{T switch} \approx \frac{2 \text{ to } 3}{\text{closed-loop bandwidth}} \cdot$$

This assumes some pretuning is applied, so that the loop error signal is only required to change VCO tuning by an amount less than the hold-in range. Under these conditions, loop behavior is more/or less linear, and reasonably well described by its closed-loop transfer function during much of the frequency-switching time. Really precise T-switch analyses or measurements are often unnecessary because the only essentiality is to ensure that the worst-case switching time is below an allowable maximum.

3. Advantages

The programmable phase-lock synthesizer (Fig. 3) has an inherently low parts count, especially when no downconverter is used. The relative simplicity and low

cost are illustrated in such synthesizers as top-of-the-line Citizens' Band transceivers and high-fidelity FM tuners. Most of the components are well suited to monolithic integrated circuits: digital counters and logic, operational amplifiers, phase
detectors, and sometimes even the VCO (with external frequency-determining components for all but the least demanding applications). Furthermore, there are few, if
any, unusual RF stray coupling and packaging problems, except keeping f_D, if any,
from contaminating the output, which is usually easily managed with one or more
isolation amplifiers. Certainly there is nothing resembling the RF isolation problems
in direct synthesizers.

Divide-and-phase-lock synthesizers are inherently capable of very wide percentage bandwidths, limited only by the VCO tuning range and K_v, achievable without sac-
rificing other desirable properties (noise and linearity). Switching among several
VCOs can easily extend the range even farther, if necessary, with only a modest pen-
alty in complexity. Hence, octave or wider bandwidths are readily provided at frequencies up to hundreds of megahertz.

Suppression of spurious sidebands and noise in phase-lock synthesizers is good
(100 dB or more) at frequencies many loop bandwidths away from the output frequency;
hence, the output is "clean" over most of its wide output bandwidth. That is a great
advantage in multichannel communications receivers and transmitters that must co-
exist in large numbers at widely varying distances, such as military aircraft and
infantry radios. Phase-lock synthesizers are common in such applications.[13] Suppression of noise and sidebands close to the carrier is more of a problem, but respectable performance (70 dB or better spurious) can readily be achieved with careful
design.

4. Disadvantages

The basic programmable phase-lock synthesizer (Fig. 3) has, in general, much
slower switching time (0.1 to 10 msec) and much coarser minimum frequency increments (1 to 100 kHz) than the other two synthesizer types discussed here. Furthermore, there is a direct and rather obvious tradeoff between these two quantities.
Fine frequency resolution is obtained by reducing f_{ref}, since adjacent output frequencies are spaced by f_{ref}. However, the maximum practical loop bandwidth is about
0.1 to 0.2 f_{ref}, so that frequency switching time is increased by reducing f_{ref}. The
noise/sideband suppression is also degraded (Design Tradeoffs, Sect. B-2).

Several fairly successful techniques get around this rather harsh tradeoff. However, switching time and frequency resolution are often inferior to that of direct or
table look-up synthesizers.

The phase detector and DC error signal circuitry that provide the VCO control
voltage (usually including an operational amplifier and lowpass loop filter, Fig. 3),
though relatively simple, must be designed and built very carefully to avoid close-in
spurious sidebands and noise, especially at the reference frequency and its harmonics.
Unless loop bandwidth is made a very small fraction of the reference frequency, which

is undesirable, there are only two ways to avoid strong reference sidebands: either a sample-and-hold phase detector must be used that produces DC output largely free of reference-frequency contamination, or else a filter that has very small phase shift in the loop bandwidth, and yet, great attenuation (70 dB or more) at the reference frequency and all harmonics must follow a multiplier or flip-flop phase detector. Neither approach is easy; both have been used successfully.

Stray audio-frequency pickup is also a problem, especially when exploiting the large percentage and absolute bandwidth capability of the phase-lock synthesis technique. This requires VCOs with wide tuning sensitivity (hence, large K_v) and large values of N. For example, consider a synthesizer with these parameters:

$$f_{out} = 100 \pm 20 \text{ MHz}$$

$$f_{ref} = 100 \text{ kHz}$$

Loop bandwidth = 10.5 kHz

$$K_L = 3.3 \times 10^4$$

$$N = 1000 \pm 200$$

$$K_v = 4 \text{ MHz per volt, or } 8\,\pi \times 10^6 \text{ radians/sec/V}$$
(5 V to tune full range)

$$K_p K_{DC} = 1.3 \text{ V/radian} \quad .$$

Now suppose 10 <u>microvolts</u> of unwanted low-frequency interference are coupled to the VCO control line (120- or 800-Hz power-line ripple). Then

$$\text{Spurious sideband level} = 20 \log_{10}\left(\frac{K_v V_n}{2K_L}\right) = 48 \text{ dB} \quad .$$

This represents fairly mediocre performance. In a typical system where this phase-lock synthesizer is surrounded by electrical and electronic equipment, every kind of stray audio-frequency coupling would have to be reduced to negligible levels to get undesired baseband signals down to one microvolt to obtain 68-dB sideband suppression. If f_{ref} were reduced to 10 kHz, and thus, K_L to 3.3×10^3, the problem is worse by a factor of 10.

In solid-state circuits, inductive coupling, nonzero ground impedances, and conducted interference on power lines are often much more troublesome than the oft-expected stray capacitance problems. The latter can cause trouble at the operational amplifier summing point , however, if it is not shielded. The operational amplifier low-frequency noise figure is obviously very important also. Such things as offset and drift can easily be traded off for lower noise in a synthesizer loop application: hence, the better FET amplifiers are very useful.

Quite often, though not invariably, the close-in phase noise performance of a phase-lock synthesizer is limited not by the noise of the reference frequency as in direct synthesis, but by the internal noise of the VCO and the achievable loop bandwidth (which determines the loop gain and the degree to which the feedback loop reduces the VCO noise). Improvements in available varactors, low-noise RF transistors and FETs in recent years have improved VCO performance, making phase-lock synthesis useful in more critical applications.

5. Augmented Divide and Phase-Lock Schemes

Several ways to avoid the harsh tradeoff mentioned in Sect. B-4 and provide fine frequency increments while retaining wide loop bandwidth, and hence, fast frequency switching time and low noise are:

a. The Digiphase technique[14] and the variations discussed by Gorski-Popiel (Chapter IV). Basically, the technique permits N to take on noninteger values by periodically increasing N by 1, and then canceling out the resulting periodic phase error.

b. Use of two or more loops separated by a fixed divider, similar to the iterative direct synthesizer.[15,16] If the f_D of a synthesizer with f_{ref} = 100 kHz is derived from the upconverted output of another similar unit that has passed through a divide-by-100 circuit, 1-kHz frequency increments occur with f_{ref} = 100 kHz in both loops.

c. Use of two loops with a Vernier technique instead of a divider.[17] If one loop has f_{ref} = 100 kHz and the other f_{ref} = 99 kHz, 1-kHz steps can be obtained in the combined output.

d. Supplying f_D from an upconverted digital table look-up synthesizer of $\pm(f_{ref}/2)$. This is a natural combination, since the two techniques have in many ways complementary advantages and disadvantages (discussed by Gorski-Popiel).

C. DIGITAL TABLE LOOK-UP SYNTHESIS

1. The Principle

The basic principle for a digital table look-up synthesizer (Fig. 4) could hardly be more fundamental: compute periodically in real time a linearly increasing phase angle, $\Theta = \omega_o t$, where ω_o is the digital input frequency command and time, t, is measured in periods of the fixed input reference frequency, f_{ref}, and then look up in a table memory the values of $\sin \Theta$ and/or $\cos \Theta$ to provide real-time output samples at the desired frequency, ω_o. The computation of Θ requires only an accumulator, i.e., an adder and a register. The register contents are increased by ω_o every cycle of f_{ref}.

Complications are introduced, primarily, to make efficient use of memory. The usual practice is to store only first-quadrant $\sin \Theta$ and add external quadrant control, address complementation, and output buffers to provide simultaneous four-quadrant sine and cosine outputs to save a factor of eight in storage compared with straight

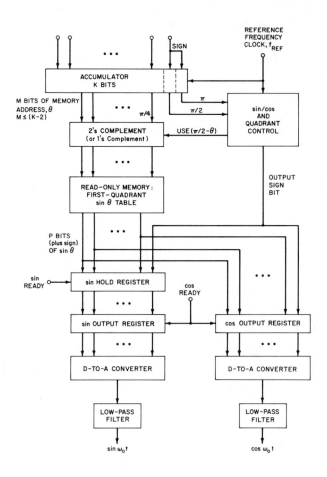

Fig. 4. Digital table look-up synthesizer.

four-quadrant sine and cosine memories. In addition, trigonometric identities are often applied within the first-quadrant memory itself to reduce the actual number of stored bits required to provide a given number, 2^M, of P-bit, first-quadrant samples.

The most useful technique yet worked out is based on splitting the memory address, Θ, into a coarse part C and a fine part F. Then:

$$\sin\Theta = \sin(C + F) = \sin C \cos F + \sin F \cos C \quad .$$

If F_{max} is sufficiently small, say, $\pi/128$ or smaller, then

$$\sin\Theta \approx \sin C + \sin F \cos C \quad .$$

The memory can then be partitioned into sections, one containing a table of $\sin C$ and the other a table of $\sin F \cos C$, the outputs being added together. Alternatively, individual sine and cosine values can be looked up and multiplied to obtain $\sin F \cos C$, saving even more memory.

A numerical example illustrates the possibilities. Suppose $M = P = 12$. A straight first-quadrant $\sin\Theta$ table memory then contains $4096 \times 12 = 49,152$ bits. However, if C is the most significant 8 bits of the overall memory address, Θ, then the $\sin C$ table contains 256 values, each 12 bits, a total of 3072 bits. The $\sin F \cos C$ table is addressed by F, the least significant four bits of M, and also by C. Only the <u>most</u> significant four bits of C are required, and only five bits of output, since the maximum value of $\sin F \cos C$ is only $(1/256)(\pi/2)$. Hence, the second memory contains 5 bits in $2^8 = 256$ addresses, a total of 1280 bits, and the total memory requirement is $3072 + 1280 = 4352$ bits, an order of magnitude saving over the straight memory. Another factor of 4 could be saved by storing fewer values of $\sin M$ and using multipliers rather than storing all products of $\sin F \cos C$.

Output registers (Fig. 4) permit the values of $\sin\Theta$ and $\cos\Theta$ to be presented simultaneously at the output, though they are extracted serially from memory. This is essential to maintain exact phase quadrature in the real-time output signals, which is important for clean single-sideband upconversion. The holding registers also provide smoothing equivalent to an analog sample-and-hold or boxcar circuit, without the difficulties of the latter. Note that aside from the inherent quantization noise, easily reduced by increasing P, output phase jitter is simply the jitter of the periodic clock labeled COS READY (Fig. 4), derived from f_{ref}. Jitter in the accumulator, memory, and control does <u>not</u> affect output noise so long as it is reasonably small compared to the sample period so that basic circuit functioning is not upset.

2. Basic Design Tradeoffs

It is reasonable and customary to provide output samples at twice the minimum Nyquist rate to make the output lowpass filters practical, i.e., four output samples per cycle at each output at the highest output frequency. Furthermore, if a bandpass output is wanted, maximum bandwidth for a given circuit speed and maximum number of frequencies for a given memory size is achieved by using both sine and cosine outputs in a single-sideband upconversion. Following these conventions (Fig. 4):

$$\text{Number of bandpass output frequencies} = 2^{K-1}$$
$$\text{Maximum } f_{out} \text{ at lowpass} = f_{ref}/4$$
$$\text{Minimum frequency increment} = f_{ref}/2^K = \text{minimum } f_{out} \text{ at lowpass}$$
$$\text{Bandpass output bandwidth} = \pm(f_{ref}/4) = (f_{ref}/2) \quad .$$

With the output samples quantized to P bits plus sign, the quantization error in the output has peak amplitude, $1/2^P$, relative to the desired signal amplitude. Spurious output suppression is thus $20 \log_{10}(2^P) = 6.02 \, P \approx 6 \, P$ dB in the worst possible case where all error energy goes into one spectral line; in practice, $6(P + 1)$ dB is a more realistic bound.[4] Having chosen P to achieve the desired output spectral purity, no more than 2^P samples of P bits each are needed in the memory to retain this performance for an arbitrarily accurate Θ. In other words, M need be no larger than P, no matter how large K, and hence, the number of output frequencies becomes the required spurious suppression upper bounds of the required memory size. However, if few frequencies are required, so that $(K - 2) < P$, then M need be no larger than $(K - 2)$ to retain P-bit output accuracy. In other words, there need never be more memory address bits than there are first-quadrant accumulator bits to address the memory. Since accumulator size is determined by the number of output frequencies, if the accumulator is small, memory size can be less than the upper bound set by the spurious requirements. This analysis is somewhat simplified in that it ignores cases where $(K - 2) > M$, but $P \neq M$, and thus blurs the distinction between error due to output sample quantization to P bits and output error due to quantization of the input address Θ to M bits, the latter error larger by $\pi/2$. However, given complete freedom, there seems little reason to make $P \neq M$ when $(K - 2) > M$; memory use would be uneconomic, since output noise is largely determined by the smaller of P or M within a few dB.

Switching time of the basic digital synthesizer is no longer than the period of the accumulator clock, $1/f_{ref}$, which can be $0.1 - 1.0$ μsec at the current state of semiconductor-memory art. The lowpass filters, however, must pass $f_{ref}/4$ and reject $3f_{ref}/4$, the alias frequency, which is only ~10 dB down at the D/A output due to the output hold. These filters will thus slow output frequency switching to roughly $4f_{ref}$, their reciprocal bandwidth; however, this still means quite respectable switching times (1-10 μsec) compared to other techniques.

3. Advantages

The digital table look-up technique is an example of direct synthesis, like iterative multistage direct synthesis and unlike phase-lock synthesis; i.e., the output frequency is produced directly by mathematical operations on the input reference frequency without internal locked oscillators or feedback loops. Hence, it is not surprising that in some ways the table look-up synthesizer is like, and even surpasses, the direct iterative synthesizer: inherently capable of extremely fast frequency switching (less than a microsecond) and almost arbitrarily fine frequency increments, hence, arbitrarily many different frequencies in a given bandwidth. The number of different frequencies increases exponentially with the amount of hardware, as with the iterative multistage synthesizer; however, the added hardware (more accumulator, and possibly, more memory bits) is so much simpler than added iterative stages that the mere number of distinct output frequencies almost ceases to be a meaningful figure of merit. One merely stops when there are enough, realizing that eventually the finest bits represent a fine adjustment for reference frequency inaccuracy.

The digital table look-up synthesizer is entirely free of RF stray coupling and shielding problems, and of microvolt-level audio-frequency pickup problems of the phase-lock synthesizer. The only packaging considerations of any importance are those of medium-speed digital circuitry (1-10 MHz clock rate; soon perhaps, 100 MHz). However, these problems are no different than for any small digital system, and a broad base of technology and experience is available in this area.

The table look-up synthesizer has the highest possible percentage of monolithic digital integrated circuits, being composed entirely of such components except for the output filters and parts of the D/A converter(s). Furthermore, the parts count is low, under a hundred packages when commercially available MSI/LSI circuits are used. Some day the entire digital section will be located on one or a few chips.

The table look-up synthesizer has one unique property: easy and arbitrary control over the initial phase of the output sinusoid at the moment of frequency switching by simply presetting the accumulator. Theoretically, this could be done with the iterative direct synthesizer by resetting all dividers in sequence, with proper allowance for and attention to the envelope delays of all those filters. Someone, somewhere, may have even tried it; however, the practical difficulties have precluded wide usefulness. Hence, the table look-up synthesizer provides a really new capability, so new that its usefulness is unexplored except for such obvious suggestions as precisely repeatable transient testing using tone-burst or frequency-stepped signals that always start at the same phase on command.

4. Disadvantages

The digital table look-up synthesizer is limited in output bandwidth by the speed capability of large-capacity (thousands of bits) semiconductor ROMs. At present, the practical limit is a few megahertz, since samples must come at a rate of about four times the maximum output frequency to permit adequate output filtering, and each sample requires one or two memory accesses plus usually other logic delays. Tens or hundreds of kilohertz bandwidth is a more confortable limit, permitting a more relaxed design and/or cheaper, low powered MOS memories. The coming of large semiconductor memories, with access times measured in nanoseconds rather than tens of nanoseconds, will presumably push the upper limit to tens of megahertz. Progress much beyond that will require essentially new memory technology, and will also tax the speed of emitter-coupled nonsaturating logic.

The output D/A converters must be absolutely first class in all respects (fast, accurate, uniform, transient-free) to preserve in the output signal the spectral purity inherent in the digital samples, which can always be improved by adding more bits. Such ideal D/A conversion is not easily achieved. Consequently, it has, to date, proved very difficult to obtain spurious output suppression better than 50 to 60 dB with any reasonable output bandwidth, especially for harmonics of the lowpass output. Such harmonics produce in-band spurious signals if selectable-sideband upconversion of sine and cosine outputs provides a nonzero center frequency, as is often required. If the lowpass output is used over only an octave bandwidth to avoid this problem, say from f_{max} to $f_{max}/2$, then the bandpass bandwidth achievable with a given lowpass output synthesizer is reduced by a factor of 4 from the $\pm f_{max}$ range achievable with selectable-sideband upconversion. This is annoying, since bandwidth is the other major problem. Furthermore, an upconversion carrier must be provided at the desired output center frequency minus $(3/4) f_{max}$, rather than at center frequency, which can be awkward.

Although selectable-sideband upconversion provides the greatest convenience and bandwidth, it must be done carefully to provide the 50-60 dB carrier and opposite-sideband suppression necessary to preserve the spectral purity achievable in the low-pass signals. Routine, off-the-shelf components and approaches yield only 30-40 dB suppression at megahertz or greater output frequencies.

References

1. R. R. Stone and H. F. Hastings, "A Novel Approach to Frequency Synthesis," Frequency, 24 (September-October 1963). See also 17th Freq. Cont. Symp., 587 (May 1963).

2. L. F. Blachowicz, "Dial Any Channel to 500 MHz," Electronics, 60 (May 2, 1966).

3. R. J. Breiding and C. Vammem, "RADA Frequency Synthesizer," Frequency 5, 25 (1967).

4. J. Tierney, C. M. Rader, B. Gold, "A Digital Frequency Synthesizer," IEEE Trans. Audio and Electroacoustics AU-19, 48 (1971).

5. N. H. Young and V. L. Johnson, "Design Principles of High Stability Frequency Synthesizers for Communications," WESCON Record, Pt. 8, 35 (W-180) (1957).

6. J. M. Shapiro and W. A. Schultz, "Development of Quartz Crystal Synthesizers," Proc. 14th Ann. Freq. Cont. Symp., 381 (31 May – 2 June 1960).

7. A. Noyes, Jr., "Coherent Decade Frequency Synthesizers," General Radio Experimenter, 11 (September 1964).

8. "A 0-50 MHz Frequency Synthesizer with Excellent Stability, Fast Switching and Fine Resolution," Hewlett-Packard 15, 1 (1964).

9. J. W. Craig, et al., "The Lincoln Experimental Terminal," Technical Report 431, Lincoln Laboratory, M.I.T. (21 May 1967), DDC AD-661577.

10. P. R. Drouilhet, "TATS – the Tactical Transmission System, Summary," Technical Note 1969-4, Lincoln Laboratory, M.I.T. (15 January 1969), DDC AD-846426. See also P. R. Drouilhet and S. L. Bernstein, "TATS – A Bandspread Modulation-Demodulation System for Multiple-Access Tactical Satellite Communication," EASCON 1969 Record, DS-8484 (14 August 1969).

11. D. G. Meyer, "An Ultra-Low Noise Direct Frequency Synthesizer," 24th Ann. Freq. Cont. Symp. (27 – 29 April 1970).

12. A. Tykulsky, "Digital Frequency Synthesizer Covering 0.1 MHz to 500 MHz in 0.1-Hz Steps," Hewlett-Packard 19 (1967).

13. R. J. Hughes and R. J. Sacha, "The LOHAP Frequency Synthesizer," 22nd Freq. Cont. Symp. (April 1968).

14. G. C. Gillette, "The Digiphase Synthesizer," 23rd Freq. Cont. Symp. (6 − 8 May 1969).

15. J. C. Shanahan, "Uniting Signal Generation and Signal Synthesis," Hewlett-Packard <u>23</u>, 2 (1971).

16. R. J. Hughes and R. J. Sacha, "A Miniature Precision Digital Frequency Synthesizer," Proc. 23rd Freq. Cont. Symp. (6 − 8 May 1969).

17. M. E. Peterson, "The Design and Performance of an Ultra-Pure, VHF Frequency Synthesizer for Use in HF Receivers," Proc. 25th Freq. Cont. Symp. (April 1971).

Chapter III
Direct Frequency Synthesis
C. Gundel

GTE Sylvania has, in recent years, been developing a capability in direct synthesizers, in conjunction with modem design and development in tactical satellite
communications systems.

These synthesizers have been characterized by successively more stringent
specifications on spectral purity and frequency switching time (defined in terms of
the phase-settling transient).

This chapter discusses design requirements, alternate approaches, implementation, and performance of a synthesizer design representative of the state-of-the-art
for frequency-hopping communications systems. The requirements for this synthesizer are 68 dB minimum (73 dB typical) discrete spurious suppression over an
approximate 13-MHz hopping range centered at approximately 111 MHz with a phase-
settling time of 2 μsec.

A. FUNCTIONAL DESCRIPTION AND CRITICAL SPECIFICATIONS

This synthesizer design (Table 1) arises from an application for very high performance, frequency-hopping, satellite communications. Actual synthesis of the
hopping spectrum occurs at VHF frequencies using appropriate frequency multiplication and translation to obtain an intermediate modem/terminal interface at UHF.
An effective multiplicative factor ($K \approx 10$) operates on the frequency-hopping range,
rms phase noise, frequency tracking errors and other frequency- and phase-related
variables in system processing of the up- and downlink synthesizer outputs.

The synthesizer generates each output frequency as a programmed, rational multiple of the instantaneous input reference frequency. This reference frequency (nom-
inally 8 MHz) is derived from a terminal-contained Doppler tracking receiver, and is
constrained to vary proportionally from its nominal value according to the ratio of
terminal/satellite Doppler velocity to the velocity of light, so that the uplink propagated frequencies and the downlink receiver local oscillator frequencies are "pre-
corrected" for Doppler offset. Under poor signal-to-noise conditions, the reference
frequency exhibits a rather poor phase-noise spectrum (Fig. 1).

The primary significance of this noisy reference tone is that frequency multiplication processes in, and subsequent to the synthesizer, produce a much degraded
phase modulation at the UHF-hopped output and UHF-dehopping downlink local oscillator (LO). An analysis of system error rate degradation has been performed,[1] ac-
counting for the effect of LO and signal noise components on M'ary frequency-shift
keyed communication link. The cited analysis proceeds from a Taylor expansion of
the theoretical error curve. Appendix A interprets that analysis, deduces the specification for the maximum allowable expectation of phase variance over a given interval, and derives an expression for this variance in terms of the phase spectrum that
is easily measurable (Table 1).

Table 1. Critical Specifications

Input

Reference frequency (average)	8 MHz ± 8 Hz
Reference frequency rate (maximum)	0.3 Hz/sec
Reference phase noise spectral purity	(defined in Fig. 1)

Output

Programmable frequency range	$\geqslant$12.5 MHz
Center frequency option	111.4 ± 0.5 MHz
Frequency programming	23-bit binary (parallel clocked TTL)
Least significant programmed increment	1.5625 Hz
Phase-settling time	2 μsec max. (to within 0.1 rad of ultimate phase)
Expected value of phase variance in maximum chip interval	1.43×10^{-4} rad^2
Discrete spurious signals	$\geqslant$73 dB (typical) below programmed tone (68-dB worst case)
Signal to total noise power	$\geqslant$57 dB (design goal-excluding 1-kHz band-centered on programmed frequency)

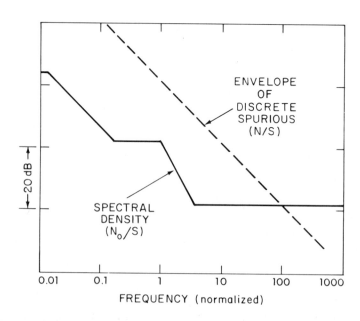

Fig. 1. Input reference (8 MHz) spectral purity. Single-sided spectral density rms noise in 1-Hz bandwidth vs baseband frequency (normalized to lowest system correlation frequency, or hopping rate).

A further loss in system performance accrues with increased phase-settling time, primarily through a loss of effective frequency chip duration, and therefore, of correlation time in the demodulator. The specification on phase-settling time, therefore, represents a budgeted sacrifice in effective chip duration at the highest chip rate.

Another critical parameter of synthesizer performance is the ratio of signal output power to total wideband noise. This specification is particularly significant in the downlink synthesizer (receiver dehopping LO). Here the effect of broadband noise in the LO of a receiver that is illuminated by a large number of downlink randomly frequency-hopping users that share a common hard-limiting satellite repeater, produce a receiver IF signal-to-noise ratio (SNR) that is approximately that of the LO SNR in that bandwidth, reduced by a factor comparable to the number of users. The specification assigned to this consideration (Table 1) is budgeted between two contributions: (a) true wideband gaussian phase noise, and (b) discrete spurious tones occurring in moderate quantity at perceptible levels and in huge quantities at extremely low levels. It is impractical to separate the second contribution from the first. Alternatively then, a specification was established arbitrarily on the allowable worst-case minimum suppression of discrete individual, spurious tones based on the assumption that the total power of all discrete spurs is unlikely to exceed the worst-case individual spur by more than 10 dB.

As the input reference tone to the synthesizer is not derived from a stable oscillator under ideal conditions, but is rather the output of a dynamic frequency tracking loop that is tracking Doppler information in a noisy environment, the possible need for simultaneously filtering phase noise contained in the input signal, while maintaining an acceptably low frequency error in tracking Doppler and acceleration related (signal) components, must be considered for the synthesizer design. These two requirements are addressed jointly in Appendix A.

B.   DESIGN APPROACH

In view of the combined requirements of (a) wide absolute frequency-hopping range, (b) rapid phase-settling time, and (c) a large number (2^{23}) of programmable frequencies, the direct synthesis approach, using multiple iterative "mix-and-divide" stages, was selected.

1.   Number of Component Frequencies

Consideration is now given to selection of the optimum number (N) of component frequencies to be made available to each stage. (This is also the frequency division factor.) The key disadvantages of the more obvious candidates for this factor (Table 2) are illustrated in the mixer spurious-product chart (Fig. 2).

2.   Values of Component Frequencies; Constraints

Having determined the design optimality of using four component frequencies, selection of values for these frequencies is required. Depicting the effect of iterative

Table 2. Selection of Number (N) of Component Frequencies

N^*	$1/N-1\ (=\alpha/\omega)$	$N/N-1\ (=\beta/\omega)$	Primary Problem(s)
2	1	2.00	Very bad in-band mixer product: $\beta = 3\omega - \alpha$
4†	0.33	1.33	Several inherent in-band products require well-balanced mixers in final mix-and-divide stages
8	0.14	1.14	LO rejects requirements for bandpass filters
16 ↓ etc.	—	—	Number of component frequencies required Decreased bandwidth of lowpass filter limits switching speed

*Only binary values of N are considered to simplify frequency control circuitry and frequency dividers. N = 3 would also involve a rather severe in-band mixer product: $\beta = 2\omega - \alpha$.

†In general, N = 4 is the logical choice, except perhaps for requirements permitting very narrow fractional bandwidths and less than maximum possible switching speeds.

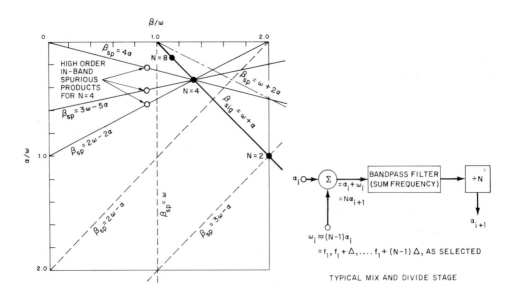

Fig. 2. Mixer product chart for N = 2, 4, 8.

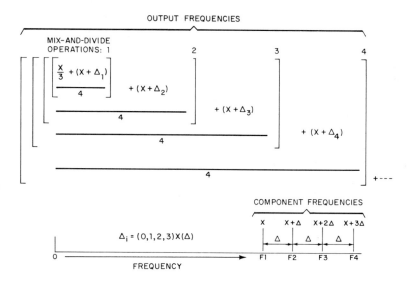

OUTPUT FREQUENCIES

Fig. 3. Effect of iterative mix-and-divide action on component frequency generation.

mix-and-divide action on frequency generation (Fig. 3), the minimum of the four equally spaced component frequencies is f_1 (set = X). The separation of adjacent component frequencies is represented by Δ. Each successively more inclusive bracket represents the output frequency of successively later and later mix-and-divide modules. The quantity $(x + \Delta_i)$ represents the magnitude of the component frequency selected at the ith mix-and-divide module. In the limiting case, where f_1 is selected at every station, the output frequency of the ith station is simply $(f_o)_i\big|_{\text{all } i} = f_1/3$. By omitting the frequency division after the final mixer, the minimum final synthesizer output frequency is extracted as $f_o\big|_{\text{min}} = 4f_1/3$, independent of the number of stages. With this condition as a point of departure and assuming n stages, an increment (increase) by Δ, 2Δ or 3Δ in the selection of a component frequency at the ith stage increases the final output frequency by, respectively,

$$\Delta(f_0) = \frac{1}{4^{n-1}} \ (\Delta, \ 2\Delta \text{ or } 3\Delta) \quad .$$

This line of reasoning yields the result for the maximum possible programmable output frequency of

$$(f_0)_{\text{max}} = \frac{4}{3} f_4 - \frac{\Delta}{4^{n-1}} \simeq \frac{4}{3} f_4 \quad .$$

The center-range programmable frequency is

$$(f_0)_{\text{mid-band}} = \frac{1}{2} [(f_0)_{\text{max}} + (f_0)_{\text{min}}] = \frac{1}{2} [\frac{4}{3} f_1 + \frac{4}{3} f_4]$$

$$= \frac{2}{3} [f_1 + f_4] = \frac{2}{3} [2f_1 + 3\Delta] = \frac{4}{3} f_1 + 2\Delta \quad . \tag{1}$$

The least equipment-programmable increment is given by

$$(\Delta f_0)_{min} = \frac{\Delta}{4^{n-1}} \quad . \tag{2}$$

Note that if the least significant bit (binary) is not exercised, then the least equipment-programmable frequency increment is

$$(\Delta f_0)_{min}\big|_{LSB \text{ fixed}} = 2\left(\frac{\Delta}{4^{n-1}}\right) \quad . \tag{2a}$$

The programmable range is given by

$$(\Delta f_0)_{max} = (f_0)_{max} - (f_0)_{min}$$

$$= \frac{4}{3}f_4 - \left(\frac{\Delta}{4^{n-1}}\right) - \frac{4}{3}f_1$$

$$= 4\Delta - \left(\frac{\Delta}{4^{n-1}}\right)$$

$$(\Delta f_0)_{max} \simeq 4\Delta \text{ (for } n \gg 1) \quad . \tag{3}$$

The design requires an exact minimum programmed output frequency increment (1.5625 Hz) and a minimum programmable range (12.5 MHz) with a nominal center frequency of 111 MHz. Using Eqs. (1), (2) and (3), suitable values can be readily determined for n, Δ and f_1 (= $f_2 - \Delta$, $f_3 - 2\Delta$, $f_4 - 3\Delta$) to provide the required center frequency, minimum programmed increment, and programmable range.

However, there exist two further constraints:

    a.   Minimum value of Δ

    b.   Maximum value of f_1 (and therefore, the maximum value of Δ for a given fractional bandwidth).

First, the bandwidth of each post-mixer bandpass filter (BPF) (that must equal approximately 4Δ as a satisfactory compromise between passing the intended mixer sum frequency and rejecting various spurious products) must be large enough to provide sufficiently low delay to meet the phase-settling time requirement. This sets a lower limit on the value of Δ. (Impact of phase-settling time on the permissible separation of component frequencies, Δ, is analyzed in Appendix B.)

To obtain high spectral purity economically from direct, iterative synthesizers above approximately 50 MHz, it is generally recommended that the synthesizer proper be designed to use component frequencies under 50 MHz with simple frequency multiplier(s) appended to obtain the final frequency band required. This is partly because crosstalk mechanisms and spurious mixer performance become troublesome, and frequency-dividers more costly and power-hungry, at higher frequencies. Direct synthesis of a centerband tone of 111 MHz requires component frequencies somewhat above 50 MHz, noting that $\Delta \ll f_1$ [Eq. (1)]. Consequently, the required frequency range and incrementation are synthesized at half value, and then frequency doubled to meet the specifications (Table 1). Thus the corresponding design specifications for the synthesizer proper (less doubler) become:

a. Programmable frequency range $\geqslant 6.25$ MHz

b. Center frequency option 55.7 ± 0.25 MHz

c. Least significant programmable increment 0.78125 Hz

   (All other specifications in Table 1 are retained as performance criteria for application at the doubled output.)

Applying Eqs. (2) and (3) to these new design bases, suitable values are deduced for n and Δ:

$$(6.25 \text{ MHz}/4) \leqslant \Delta = 4^{n-1} \times 0.78125 \text{ Hz} \quad , \tag{4}$$

giving:

$$n\Big|_{\text{an integer}} \geqslant 1 + \log_4 [(6.25 \times 10^6)/(4 \times 0.78125)]$$

$$n \geqslant 1 + \log_4 (2 \times 10^6)$$

$$\geqslant 1 + (\log_4 10)(\log_{10} 2 \times 10^6)$$

$$\geqslant 1 + (1.66)(6.301)$$

$$n \geqslant 1 + 10.46$$

$$n \geqslant 11.46$$

yielding

$$n_{\text{min}} = 12 \text{ stages} \ (= 11 \text{ mix-and-divide stages} + 1 \text{ mix-only stage})$$

$$\Delta = 4^{11} (0.78125 \text{ Hz})$$

$$\Delta = 3.2768 \text{ MHz} \quad . \tag{5}$$

Equation (5) is a valid solution in terms of switching speed, but requires a programmable range [Eq. (3)] of 13.1072 MHz at each mixer output that is more than twice the required (undoubled) tuning range (6.25 MHz). In such a situation, it is appropriate, in the interests of easing the design problems of the iterative stage BPFs, to halve the absolute (and fractional) bandwidths. This can be done by fixing the command to the least significant frequency control point (see Fig. 4, particularly, M/D 1).

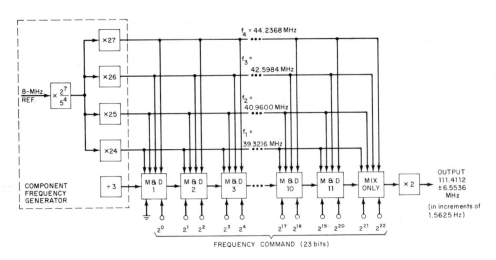

Fig. 4. Block diagram of synthesizer.

Using Eq. (2a) instead of Eq. (2), a new value of Δ is obtained,

$$\Delta = 3.2768 \text{ MHz}/2 = 1.6384 \text{ MHz} (>990 \text{ kHz})^* \qquad . \tag{6}$$

Now, from Eq. (1),

$$\text{Center frequency} = 55.7 \pm 0.25 \text{ MHz} = \frac{4}{3} f_1 + 2\Delta \tag{7}$$

and f_1 (as well as f_2, f_3 and f_4) must be simple rational multiples of 8 MHz, the reference input to the synthesizer. Whence

$$f_1 = \frac{3}{4} [(55.7 \pm 0.25) - 3.2768] \text{ MHz} \qquad ,$$

requiring

$$39.3174 \leqslant f_1 \leqslant 39.5049 \text{ MHz} \qquad .$$

To simplify the generation of f_2, f_3 and f_4, f_1 should also be picked as a simple rational multiple of ($\Delta = 1.6384$ MHz), if possible.

As an acceptably simple rational multiple of 8 MHz, and (for this particular design), as an integer multiple of ($\Delta = 1.6384$ MHz),

$$\Delta = (8 \text{ MHz}) (2^7/10^2) (2^2/5^2) = 1.6384 \text{ MHz} \tag{8}$$

$$f_1 = (8 \text{ MHz}) (2^7/10^2) (2^2/5^2) (24) = 39.3216 \text{ MHz} \tag{9}$$

$$(\leqslant 39.5074,$$

$$\geqslant 39.3174 \text{ MHz})$$

---

* Derivation of requirement for minimum permissible value of Δ (Appendix B).

similarly,

$$f_2 = (8 \text{ MHz}) (2^7/10^2) (2^2/5^2) (25) = 40.9600 \text{ MHz} \tag{10}$$

$$f_3 = (8 \text{ MHz}) (2^7/10^2) (2^2/5^2) (26) = 42.5984 \text{ MHz} \tag{11}$$

$$f_4 = (8 \text{ MHz}) (2^7/10^2) (2^2/5^2) (27) = 44.2368 \text{ MHz} \quad . \tag{12}$$

The design thus far is reflected in Fig. 4.

3.   Implementation of Component Frequency Generator

The next step in the design is to determine how best to implement the component frequency generator.   Three general approaches are available:

a.   Because of the integral multiple relationship between the component frequency separation Δ, and the component frequencies themselves, the latter could be extracted from a harmonic spectrum of Δ by individual filters.

b.   Inasmuch as only four higher order harmonics of Δ are required, they could be extracted by filters identical to those required for the first approach, from the output to a modulator multiplying $[(51/2) (\Delta)]$ by $(\pm\Delta/2, \pm3 \Delta/2)$.

c.   Synthesize each component frequency in an individual phase-locked loop (PLL), taking advantage of the common factors in Eqs. (8) through (12) to simplify the circuitry.

The third approach, although the most expensive of the three, is necessary in this application to (a) obtain required mutual isolation between component frequencies (about 80 dB at 44 MHz) (crystal filters with this capability are not really practical), and (b) adequately filter the low frequency noise components in the input reference tone spectrum.

Synthesizer design must assure sufficiently low output phase noise to meet its budgeted maximum contribution to system error rate degradation.   Simultaneously, sufficient tracking capability must be provided to keep output frequency error perturbations (and resultant phase error rates) due to abrupt terminal acceleration (±1 G) sufficiently small to contribute comparably low error rate degradation.

The input reference phase noise spectral density (Fig. 1) is multiplied by the frequency multiplication factor $(111/8)^2$ between the synthesizer input and output terminals [Fig. 5(a)].   Applying the appropriate weighting function [derived in Appendix A, and depicted in Fig. 5(b)] an effective output noise density function is obtained [Fig. 5(c)] that can be integrated to obtain the worst-case expected phase variance (during the longest correlation time, or frequency-hopping chip interval).   With no spectral filtering provided by the synthesizer, the resultant phase variance due to reference tone noise is excessive:

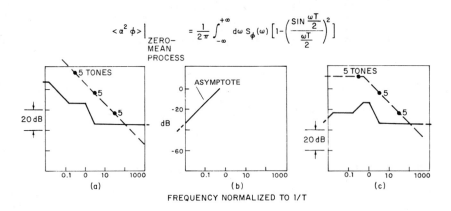

$$\langle a^2 \phi \rangle \Big|_{\substack{\text{ZERO-}\\\text{MEAN}\\\text{PROCESS}}} = \frac{1}{2\pi} \int_{-\infty}^{+\infty} d\omega \, S_\phi(\omega) \left[1 - \left(\frac{\text{SIN}\frac{\omega T}{2}}{\frac{\omega T}{2}} \right)^2 \right]$$

FREQUENCY NORMALIZED TO 1/T

Fig. 5. Spectral considerations of phase noise (comb filter implementation of component frequency generation): (a) input reference phase noise (discrete tone envelope +N_0/S), (b) spectral weighting function appropriate for correlation time, T, and (c) product of weighting function (5b) and spectra (5a) (=effective spectra for correlation time, T).

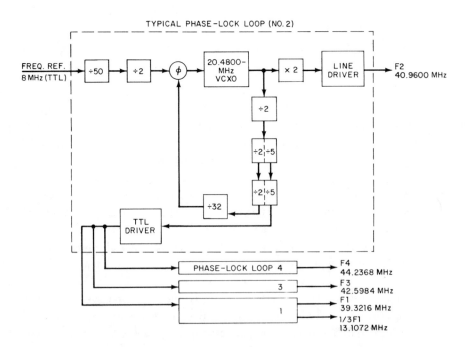

Fig. 6. Block diagram of component frequency generator (PLL implementation of component frequency generation).

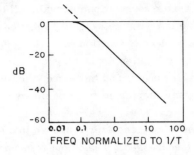

Fig. 7. Phase transfer function of PLL implementation of component frequency generator.

Nonfiltering synthesizer:

$$\sigma^2(\phi_0)\big|_{\text{ref. noise}} = (0.93 \times 10^{-4})\ \text{rad}^2 \approx (1.43 \times 10^{-4})\ \text{rad}^2 \qquad (13)$$

$$\approx \text{max. specification} \quad .$$

Therefore, the component frequency generator is implemented using the PLL approach (Fig. 6) with a resultant lowpass bandwidth (Fig. 7). Although throughput reference noise power is reduced by band-limiting, the PLLs actually contribute to phase noise density at the synthesizer output [Fig. 8(a)]. By weighting this net phase noise and integrating the phase variance during the maximum correlation interval, the specification (Table 1) is met.

Filtering synthesizer:

$$\sigma^2(\phi_0)\big|_{\text{ref. noise}} = (0.22 \times 10^{-4})\ \text{rad}^2 \ll (1.43 \times 10^{-4})\ \text{rad}^2 \qquad (14)$$

$$\ll \text{max. specification} \quad .$$

[To perform the Eq. (14) analysis with respect to meeting the same phase variance specification applied to any higher data rate (higher frequency-hopping rate and

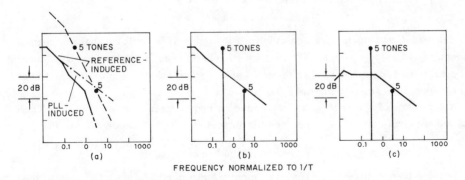

Fig. 8. Spectral consideration of phase noise (PLL implementation of component frequency generation): (a) components of reference-induced PLL-induced synthesizer phase noise (nonweighted), (b) total synthesizer phase noise (nonweighted), and (c) effective spectra for correlation time, T (7b, weighted by curve 5b).

inversely shorter correlation interval), all nonweighted curves (Figs. 5, 8) should be shifted to the left with the weighting function fixed. Inasmuch as the filtered, non-weighted noise density function essentially decreases monotonically, the weighted, filtered density function for any higher data rate will be lower in magnitude for any normalized frequency; and therefore, the integrated value of expected phase variance during any shorter correlation interval will be less than that computed for the lowest data rate.]

The dynamic synthesizer response to a sustained terminal acceleration of ±1 G must now be addressed. The output frequency (at the 111-MHz interface) must not suffer an offset that contributes a phase power component that, when added to the left member of inequality Eq. (14), produces total power in excess of the specification.

Considering a PLL, in a servo sense, for a single PLL in cascade with any combination of frequency multiplicative or divisive processes that are relatively wideband, the fractional frequency offset can be expressed:

$$\Delta f_0/f_0 = (\ddot{X}/C)(1/K_v) \tag{15}$$

where

$$\ddot{X} = \text{relative acceleration between terminal and satellite (m/sec}^2)$$

$$C = \text{velocity of light (m/sec)}$$

$$K_v = \text{"velocity error constant" of PLL (rad/sec)} \quad .$$

A cascade of two PLLs occurs in this design approach, inasmuch as the loops to generate f_1, f_3 and f_4 are supplied from the feedback chain (outputs) of the f_2 PLL. For the case of n cascaded PLLs, the fractional error in output frequency is given by

$$\Delta f_0/f_0 = \sum_{i=1}^{n} \ddot{X}/C \, (1/K_{v_i}) = \ddot{X}/C \sum_{i=1}^{n} (1/K_{v_i}) \quad . \tag{16}$$

Representative normalized calculations in Appendix A indicate that pertinent properties of the loop filters, K_v, in each PLL have readily been controlled, adding acceptably small contributions to the total phase noise power over the longest correlation interval prescribed for the system.

A summary of synthesizer performance vs requirements is presented in Table 3. Detailed analysis supporting these results is included in Appendices A and B.

4.   Wideband Noise vs Discrete Spurious Tone Tradeoff

In the mix-and-divide synthesizer, most wideband noise is attributable to the amplifier following the final mixer. As the final mixer contributes the major portion of the in-band spurious mixer product, which worsens rapidly at increased mixer output levels, a compromise between suppressing the mixer products (discrete spurs) and suppressing the noise floor at the synthesizer output, is required. Both of these requirements are met by specifying a good doubly balanced mixer in the final slot,

Table 3. Critical Output Specifications

Parameter	Design Basis	Performance
Phase-settling time	2 µsec	1.5 µsec
Average phase noise power in maximum chip interval	1.43×10^{-4} rad^2	0.60×10^{-4} rad^2 (two terminals with 2 G relative acceleration)
Discrete spurious suppression	73 dB, typical 68 dB, minimum	75 dB, typical 68 dB, minimum
Signal to total noise power (excluding 1-kHz band centered on programmed frequency)	57 dB, design goal 50 dB, minimum	53.5 dB

operating it at an input signal level that optimized the in-band products (Fig. 2), and matching the post-mixer BPF in a low-noise amplifier.

5.   Crosstalk and Isolation Problems

Crosstalk and isolation problems arise primarily because of the need for compact packaging.

a.   Component Frequency Crosstalk

Since the synthesizer's operating mode involves successive selection and processing of one or another of four nearly equal basic component frequencies, it is apparent that if the intended component frequency is contaminated at any point in the chain with one or more of the three unintended component frequencies, in-band sidebands are generated and propagated. Such propagation is accompanied by improvement in signal-to-spurious-sideband-ratio of 12 dB at each frequency-division point if hard-limiting is provided in each stage to remove amplitude modulation components. (Another important benefit of hard-limiting with plenty of overdrive, is the avoidance of "dead zones" in frequency-divider action during the "moment of energy transfer" in the BPFs during the frequency switching transient.)

The first legitimate opportunity for crosstalk occurs on the multilayer motherboard to which all subassemblies communicate by soldered wire connection, including the component frequency "transmission lines." Signals are extracted as required from these "transmission lines" into 12 mix (/divide) subassemblies, each a sealed, shielded unit spaced along the length of the motherboard. Only one of the four component frequencies is accepted into each subassembly, the choices being independent of one another and uniquely defined by the ultimate frequency required from the synthesizer.

Crosstalk on the motherboard is minimized by floating each line with respect to the motherboard ground at the source and load end via transformers. The load-end transformers must be wound uniformly to minimize crosstalk in the switch circuitry of the mix-and-divide subassemblies. (Particularly troublesome crosstalk problems have been experienced in synthesizers using a "spectrum" generator type of component-frequency generator plus comb filter.)

b. Other Crosstalk Problem Areas

The need to avoid leakage of high level signals, and to confine stray fields and define ground paths, is extremely important in a synthesizer of this type. For example, the output of a divide-by-four circuit, whose input signal frequency f is accompanied by spurious signal f_n, contains a great number of spurious frequencies $[(n/4) f_s \pm f_n]$. Such a prolific family of spurious frequencies has an excellent chance of including immediately, or producing subsequently, an in-band spurious product. This is just one example of how spurious products result from "indirect" causes .

6. Summary of Key Points

A rationale is presented:

a. For selecting a number of component frequencies for a given design situation in direct synthesis, and

b. As a guide for the optimum selection of values of component frequencies constrained by transient response (phase-settling behavior) and discrete spurious suppression requirements (isolation problems vs center frequency).

Synthesizer design-related contributions to the degradation of error rate performance is evaluated for two implementations of the basic design, one using phase-locked oscillators for component-frequency generation, the other using a spectrum generator and comb filter. In both cases, the synthesizers control up- and downlink frequency hopping in a M'ary MFSK modulation situation.

Appendix A

Evaluation of System Performance vs Synthesizer

Phase Noise and Tracking Error

These calculations are based on the interpretation that the appropriate noise power representation used to analyze error rate degradation, in particular, the resultant expression (Taylor expansion of theoretical error curve cited by T. Seay),[1] is the expected value of the power in the phase error function during the correlation interval, T, taking into account all phase noise contribution, both up- and downlink. Thus

$$(\Delta P_E)_{max} = 1.7 \times 10^{-2} \, [\Sigma(\text{all phase noise component powers})] \qquad (A\text{-}1)$$

where P_E is taken to be 10^{-3}, a nominal probability of bit error assumed in the absence of any phase noise.

The phase noise power, attributable to synthesizer implementation, is considered to consist of three additive portions: two due to random phase noise and low frequency discrete spurious components (both zero-mean processes, by virtue of the phase-locked nature of the synthesizer), the third being due to an average frequency offset, sustained over the correlation interval, due to dynamic tracking error only. (The frequency error contribution due to finite incrementation resolution is <u>not</u> considered as a contribution attributable to synthesizer implementation, but <u>is</u> assumed to be in the system budget.)

The literature[2] offers an expression to evaluate the average variance of a zero-mean signal, $\phi(t)$, observed (repeatedly) for a finite time interval of duration, T, in terms of the spectral density, $S_\phi(\omega)$, of that signal. This result is

$$\langle \sigma^2(\phi) \rangle \Big|_{\substack{\text{zero-}\\ \text{mean}\\ \text{process}}} = \frac{1}{2}\, \pi \int_{-\infty}^{+\infty} d\omega \, S_\phi(\omega) \left\{ 1 - \frac{\sin^2(\frac{\omega T}{2})}{(\frac{\omega T}{2})^2} \right\} \qquad (A\text{-}2)$$

showing that the effect of limiting each independent observation interval to T seconds is to reduce the influence of the low frequency portion of the spectrum toward contributing to the expected variance seen within that interval. To determine the variance, therefore, the phase power spectral density function is weighted by the multiplicative factor $[1 - \sin^2(\omega T/2)/(\omega T/2)^2]$ before integrating (over all frequencies). As Cutler and Searle[2] point out, this is equivalent to observing the true spectral density function through a high-pass filter with asymptotic limbs:

$$\frac{1}{3}\,(\omega T/2)^2 \text{ for } \omega T/2 \ll 1$$

$$1 \text{ for } \omega T/2 \gg 1 \quad ,$$

and cutoff frequency, $\omega_c \approx 2/T$.

In summary then, the average variance in synthesizer output phase may be evaluated by integrating the appropriately weighted phase noise density function. If the up- and downlink synthesizers contribute equally (and independently) in this respect, then the net phase variance contributing to system error rate degradation (due to random causes) must be taken as twice that of either synthesizer alone.

With respect to the appropriate treatment of a possible frequency offset maintained throughout the correlation interval of the demodulator (in response to a ramp in input reference frequency), the resultant phase ramp can be considered to impact correlation only to the extent that the phase differs from its average value within interval T. Accordingly, it may be determined that the resultant net contribution to phase power within interval T is given by

$$\sigma^2(\phi_T)\Big|\text{due to frequency offset} \triangleq (\phi_P)^2/12 \tag{A-3}$$

where ϕ_P is defined as the peak deviation in phase from its initial value in interval T. Now

$$\phi_P = T(2\pi \, \Delta f_{syst}) \text{ radians}$$

$$= T(2\pi)(2K)(\Delta f_{synth}) \text{ radians}$$

where

$$\Delta f_{syst} = \text{frequency offset at system usage points}$$

$$\Delta f_{synth} = \text{frequency offset at individual synthesizer output}$$

$$K = \text{frequency multiplicative factor } (\sim 10).$$

Therefore,

$$\sigma^2(\phi_T)\Big|_{\text{frequency offset}} = \frac{1}{12}(2\pi T \Delta f_{syst})^2$$

giving

$$\Delta f_{syst} = \frac{1}{\pi T}\sqrt{3\sigma^2(\phi_T)}\ \Big|\begin{array}{l} = 2K\Delta f_{synth} \\ \text{additive} \\ \text{frequency offset} \end{array} \tag{A-4}$$

Solving for the phase noise power component attributable ultimately to the limited ability of the component-frequency generator PLLs to track the acceleration-induced ramp in the input reference frequency,

$$\sigma^2(\phi_T)\Big|_{\substack{\text{frequency} \\ \text{offset}}} = \frac{1}{12}[2\pi(2K)\,T\Delta f_{synth}]^2 \text{ rad}^2 \quad . \tag{A-5}$$

Substituting from Eq. (16) for the frequency offset ($f_{synth} \triangleq f_o$),

$$\sigma^2(\phi_T)\bigg|_{\substack{\text{frequency} \\ \text{offset}}} = \frac{1}{12}\left\{4\pi KTf_{\text{synth}}\sum_{i=1}^{n}\frac{1}{K_{v_i}}\frac{\ddot{X}}{C}\right\}^2 \text{rad}^2 \quad . \tag{A-6}$$

In words then, Eq. (A-6) expresses the acceleration-induced phase noise power at the frequency-multiplied system usage interface in terms of the:

> frequency multiplication factor, K
>
> correlation time interval, T
>
> (non-multiplied) synthesizer output frequency, f_{synth}
>
> ($n = 2$) cascaded PLL "velocity error" constants, K_v
>
> terminal/satellite accelerations, each assumed $= \ddot{X}$
>
> velocity of light, C

Now the maximum permissible phase noise power of Eq. (A-1) must be budgeted among its components:

$$\sum\substack{\text{component} \\ \text{phase noise}} = \sigma^2(\phi_T)\bigg|_{\substack{\text{Additive} \\ \text{frequency offset}}} + \sigma^2(\phi_T)\bigg|_{\substack{\text{random uplink} \\ \text{noise}}}$$

$$+ \sigma^2(\phi_T)\bigg|_{\substack{\text{random down-} \\ \text{link noise}}} + \sigma^2(\phi_T)\bigg|_{\substack{\text{uplink discrete} \\ \text{(LF) spurs}}}$$

$$+ \sigma^2(\phi_T)\bigg|_{\substack{\text{downlink discrete} \\ \text{(LF) spurs}}} \quad . \tag{A-7}$$

(In relating system performance to various phase noise terms, it was necessary to reference the frequency-multiplied interface. The performance/specification relationships as referenced to the synthesizer output terminals are discussed next.)

Allowing a degradation in bit error rate (from a baseline of 1×10^{-3}) due to all synthesizer implementation-related phase noise sources corresponding to a specific small fraction of 1-dB loss in system sensitivity, Eqs. (A-7) and (A-1) yield the requirement:

$$\sum\substack{\text{component} \\ \text{phase noise} \\ \text{powers}}\bigg|_{\substack{\text{referenced to} \\ \text{each individual} \\ \text{synthesizer output}}} \leqslant 1.43 \times 10^{-4} \text{ radians}^2 \quad . \tag{A-8}$$

Projecting no (PLL) lowpass filtering of the synthesizer output phase noise (random plus discrete) of the reference signal, the first term (frequency offset effect) of the right member of Eq. (A-7) is negligible. Integration of the weighted density functions yields a net random component of phase variance of (0.12×10^{-4}) rad^2. Further, if the (weighted) discrete component envelope exists, at five tones per frequency decade (15 tones total), the net phase variance attributable to these components (assumed uncorrelated) is computed as (0.88×10^{-4}) radians^2.

The total phase variance over interval T for the non-PLL implementation is, therefore,

$$\sigma^2(\phi_T)\Big|_{\substack{\text{total} \\ \text{(non-PLL)}}} = (1.00 \times 10^{-4}) \text{ radians}^2 \quad ,$$

which is unsatisfactory with respect to the requirement of Eq.(A-8) (too marginal).

Similar calculations for the PLL implementation yield the following (acceptable) phase power components (for each synthesizer, one uplink related, one downlink related):

$$\sigma^2(\phi_T)\Big|_{\substack{\text{add frequency} \\ \text{offset}}} = 0.38 \times 10^{-4} \text{ radians}^2$$

$$\sigma^2(\phi_T)\Big|_{\text{random}} = 0.03 \times 10^{-4}$$

$$\sigma^2(\phi_T)\Big|_{\text{discrete}} = 0.19 \times 10^{-4} \quad .$$

Therefore, adding these components,

$$\sum \begin{array}{l} \text{component} \\ \text{phase noise} \\ \text{powers} \end{array} \Bigg|_{\substack{\text{total} \\ \text{(PLL)}}} = 0.60 \times 10^{-4} \text{ radians}^2 \quad .$$

The phase noise power obtained for the PLL implementation (0.60×10^{-4} radians^2) is more acceptable than the corresponding variance of the comb filter approach (1.00×10^{-4} radians^2) because the former value occurs only under a very improbable and highly transitory combination of terminal dynamics, whereas the latter variance contributes a constant degradation in system error rate from random noise and equipment-related spurious effects.

Appendix B
Impact of Phase-settling Time Requirement on Minimum
Permissible Separation (Δ) of Component Frequencies

In considering the transient response of a direct synthesizer to a command change frequency, it is recognized that a lower limit exists on the allowable fractional programmable bandwidth at a given center frequency within the iterative mix-and-divide structure, consistent with a specific frequency and/or phase-settling time requirement.

For any tentative selection of fractional bandwidth and number of component frequencies, the attendant specification on discrete spurious suppression forces a selection of BPF type, bandwidth, and number of poles. The phase-settling behavior of the resultant synthesizer implementation can be modeled closely in terms of the envelope delay, t_d, of the individual stage filter by considering the effect of a frequency increment effected at any given stage as delayed by the sum of the envelope delays of all dominant BPFs intervening between the point of execution and final synthesizer output. However, each such delayed (partial) output frequency increment appears reduced in magnitude by the net frequency division factor. The effect of superimposing such idealized (frequency × time) blocks in terms of the corresponding rate of approach of the composite output phase to its asymptotic value is shown in Fig. B-1.

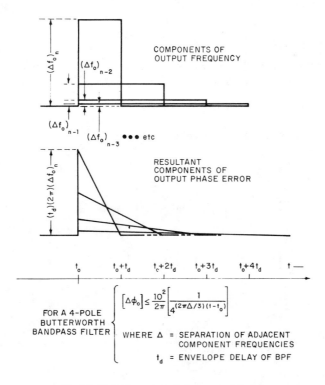

Fig. B-1. Model for analysis of phase-setting behavior.

It can be determined by examination (Fig. B-1) that the net phase error residual at time $(t = t_0 + mt_d)$ will not exceed

$$\Delta\phi_0\Big|_{max} = (t_d)\,(2\pi)\,(\Delta f_{max})\left(\frac{1}{4^m}\right)[1 + \frac{2}{4} + \frac{3}{16} + \frac{4}{64} + \dots]$$

or

$$\Delta\phi_0\Big|_{max} = 2\pi t_d\,\Delta f_{max}\left[\frac{1.78}{4^m}\right] = (33.6\,\Delta)\left[\frac{t_d}{4^m}\right] \qquad\qquad (B-1)$$

where

$\Delta\phi_0$ = output phase error, due to all (n) stages ... (radians)

$(\Delta f_0)_i$ = frequency selection increment commanded at the ith

stage ... (Hz)

Δf_{max} = maximum possible value of $(\Delta f_0)_i$ = 3Δ ... (Hz) .

Now, assuming (as is consistent for the combination of fractional synthesizer hopping bandspread and the four component frequencies in this design) that the dominant BPF following each stage mixer is a 4-pole Butterworth design, then the envelope delay may be estimated from available transient response curves as

$$t_d\Big|_{\text{4-pole Butterworth}} \approx [\frac{6}{\pi(BW)}] \quad .$$

Also, the bandwidth (BW) of this filter, to provide adequate suppression of "LO" and near-band spurious tones, must fulfill the requirement

$$4\Delta \approx BW\ (\approx 6/\pi t_d,\ \text{from above}) \quad .$$

Therefore, substituting for t_d, where:

$$t_d = t - t_0$$

$$(\Delta\phi_0)_{max} \approx (33.6\Delta)\,(\frac{3}{2\pi\Delta})\left\{\frac{1}{\left[\frac{(t-t_0)\,2\pi\Delta}{3}\right]}\right\}$$

$$\approx \frac{10^2}{2\pi}\left[\frac{1}{\frac{2\pi\Delta}{3}(t-t_0)}\right]\text{radians} \quad .$$

Solving for Δ,

$$\frac{2\pi}{3}\,(\Delta)\,(t-t_0) \approx \log_4\left[\frac{10^2}{2\pi(\Delta\phi_0)_{max}}\right] \quad ,$$

giving, as a requirement for minimum component frequency spacing, consistent with a post-mixer filter BW adequate to provide acceptably low phase error $(\Delta\phi_0)_{max}$ within a settling time $(t - t_0)$:

$$\Delta = \left[\frac{3}{2\pi(t - t_0)}\right] \log_4 \left[\frac{10^2}{2\pi(\Delta\phi_0)_{max}}\right]$$

$$= \frac{3}{2\pi} \left[\frac{1}{(t - t_0)}\right] \left[\log_4 \frac{10^2}{2\pi} - \log_4 (\Delta\phi_0)_{max}\right]$$

$$\Delta \approx \frac{3}{2\pi} \left[\frac{1}{(t - t_0)}\right] [2.0 - 1.66 \log_{10} (\Delta\phi_0)_{max}]$$

where

Δ is in Hz

$mt_d = t - t_0$ (sec)

$(\Delta\phi_0)_{max}$ is in radians .

Now the synthesizer phase-settling specifications stipulate that

$$\Delta\phi_0 \leqslant 0.05 \text{ radians at } t = t_0 + 2 \text{ } \mu\text{sec}$$

giving

$$\Delta \geqslant \frac{3}{2\pi} \left[\frac{1}{2 \times 10^{-6}}\right] 2.0 - 1.66 \log_{10} (0.05)$$

$$\Delta \geqslant \frac{3}{2\pi} \left[\frac{4.16}{2 \times 10^{-6}}\right] = 9.9 \times 10^5 \text{ Hz} .$$

Thus the spacing, Δ, of component frequencies must exceed 990 kHz, to assure meeting the phase-settling time specification.

References

1. T. Seay and S. Bernstein, private communication.

2. L. S. Cutler and C. L. Searle, "Some Aspects of the Theory and Measurement of Frequency Fluctuations in Frequency Standards," Proc. IEEE 54, 2 (February 1966).

   The following references are recommended for general information in the titled areas:

   E. J. Baghdady, R. N. Lincoln and B. D. Nelin, "Short-term Frequency Stability: Characterization, Theory and Measurement," IEEE/NASA Symp. on Short-term Freq. Stab., Goddard Space Flight Center, Greenbelt, Md. (November 23-24, 1964).

   D. G. Meyer, An Ultra-low Noise Direct Frequency Synthesizer, John Fluke Mfg. Co., Inc., Seattle, Washington.

Chapter IV
Phase-Locked Loop Frequency Synthesizers
J. Gorski-Popiel

The purpose of this chapter is to help the reader "understand" phase-locked loop (PLL) synthesizers. A PLL in its most fundamental form is a simple servo-system (Fig. 1). Signal, S_2, out of an oscillator is compared with reference signal, S_1, producing an error, E, that is applied to the oscillator. The effect of the loop is to make E = 0, and hence, S_2 can be made to bear a fixed relationship to S_1.

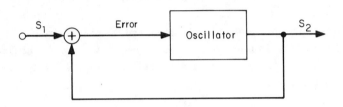

Fig. 1. Servo loop.

This principle is used for frequency synthesis in many ways. However, instead of discussing existing synthesizers, a detailed discussion of a design capable of achieving state-of-the-art performance (Table 1) should present the general concepts involved in PLL design.

Table 1. Feasible Synthesizer Specifications

Bandwidth	200-300 MHz max
Frequency spacings	Down to fractions of Hz
Switching time	200 μsec min
Frequency stability	1×10^9
Spurious suppression	>70 dB

A. SYNTHESIZER TYPES

There are two basic methods of frequency synthesis. The conventional and most straightforward method is to generate a spectrum from a frequency standard and then extract the required frequency using mixers, dividers and filters. Accordingly, this method is sometimes referred to as "direct synthesis."

The other approach is to use a PLL consisting basically of a voltage-controlled oscillator (VCO), whose output is divided by N and multiplied by an incoming reference signal. The resultant signal is then lowpass filtered and used to control the VCO. The effect of the loop is to make the error voltage go to zero (Fig. 2). When this occurs the VCO frequency and phase are N times those of the reference signal. Thus, by suitable choice of N, any required frequency can be obtained.

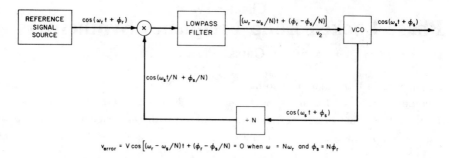

Fig. 2. Basic phase-locked loop.

Each of these approaches has its advantages and disadvantages (Table 2). Although performance of the PLL synthesizer is poorer in some ways than the direct method, its potentially small size and power consumption make it attractive for applications where these are at a premium.

Table 2. Basic Synthesizer Properties Compared

Direct Synthesis	PLL Synthesis
High resolution possible (fractions of 1 Hz)	High resolution possible (fractions of 1 Hz)
High switching speed (few μsec)	Moderate switching speed (200 μsec)
Good stability (1 in 10^9)	Good stability (1 in 10^9)
Relatively high power consumption	Low power consumption
Large size	Small size
Complexity proportional to hopping bandwidth	Good spurious suppression
Prone to RF interference	

B. PLL SYNTHESIZER FUNDAMENTALS

Characteristic parameters for digitally controlled PLL synthesizers are bracketed in Fig. 3. The VCO output signal is fed into a counter whose capacity is adjusted externally to N. The counter generates one count for every cycle of the incoming frequency. So after N counts the counter overflows, returns to zero, and starts counting all over again. The period of this overflow is N times larger than that of the incoming VCO frequency, i.e., overflow frequency = f_v/N. The phase of the counter output is ϕ_v/N. Let the overflow period be T. Since any change in f_v can be observed by the system only after the counter overflows, i.e., T sec later, the effect of the counter is to also produce a delay of up to T sec.

The phase detector can be modeled in its simplest form by an ideal multiplier. The reference wave with frequency, f_R, and the counter overflow wave with frequency, f_v/N, are multiplied in the phase detectors. As the lowpass filter eliminates the upper sideband the VCO input voltage, v_2, varies with the difference frequency,

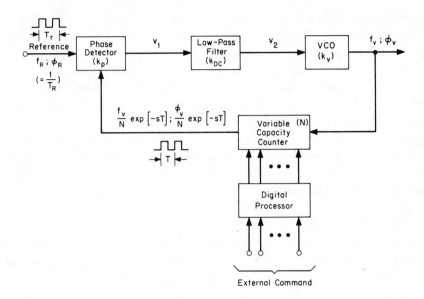

Fig. 3. **Digitally controlled phase-locked loop synthesizer.**

$f_R - f_v/N$. This, in turn, produces a swept frequency output, since $f_v = k_v v_2$. The system achieves equilibrium when the VCO frequency stabilizes to one value. This can happen only when v_2 becomes a DC level, i.e., when the loop has reached an equilibrium state and the VCO frequency, $f_v = N f_R$. Under these conditions, the VCO is said to be locked to the reference. Any variations in the reference appear multiplied N-fold on the VCO signal. The reference, therefore, must be as clean and as constant as possible. Fortunately, it is possible to achieve extremely pure signals of high stability at a single frequency. It is impossible to duplicate that purity in a variable wideband oscillator. Reference oscillators can usually be constructed with such signal stability that even when multiplied by an N of 1000 or more, the stability exceeds that obtainable from a VCO in an open-loop mode. In a PLL, the VCO frequency and phase are N times that of the reference. Since N can be varied externally over the full range of the VCO, a PLL offers a way to achieve high stability and wide bandwidth.

Since N has to be an integer, the smallest frequency difference between consecutive permissible frequencies out of the VCO is $f_R (f_v = N f_R)$. Hence, the reference frequency has to be equal to the smallest step. Unfortunately, if the resolution is to be less than several tens of kilohertz, the synthesizer settling time may become excessively long, and the sensitivity of VCO frequency to mechanical vibration may become important.

Settling time can be defined as the time it takes the VCO to transfer 90 percent of the energy to a new frequency once the command to change frequency is given. This time to a first approximation is directly proportional to the reference period, T_R. Typically, 20 to 30 reference periods are required. If, for example, $f_R = 10$ kHz, then T_R = 100 μsec and the settling time will be between 2 and 3 msec. If finer frequency spacing and shorter settling times are required, the basic simple loop becomes inadequate and a more sophisticated system has to be used.

The foregoing holds only for VCOs that do not introduce time lags of their own. In practice, most VCOs do have finite response times. The most severe problem in this area is the thermal lag in the varactor diode of the VCO. Great care has to be taken in the manufacture of VCOs to minimize these effects.[1,2]

Two quite different schemes to achieve high resolution have been investigated: one based on the "Digiphase" principle, the other a straightforward combination – via single sideband (SSB) mixer – of a simple PLL synthesizer with coarse resolution and an interpolation synthesizer with fine resolution.

1. Digiphase Principle

A paper published in Frequency Technology, August 1969, by Gerry C. Gillette describes the Digiphase principle and provides some circuits. The principle can be explained by use of the version shown in Fig. 4.

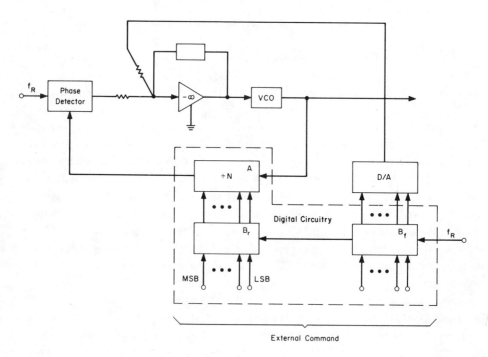

Fig. 4. Digitally controlled PLL with high frequency resolution.

It was demonstrated in the preceding section that the VCO frequency, f_V, in a simple PLL is related to the reference frequency, f_R, and the divide ratio, N, by

$$f_V = Nf_R \quad . \tag{1}$$

If N can be made to assume fractional values, the frequency resolution problem is solved. For example, if N can be incremented in fractional steps, the frequency resolution of f_v can be made arbitrarily small. Since a digital counter cannot provide a fractional count, a way has to be found to modify the counter's overflow pattern externally to effectuate a fractional count. The most straightforward way is to delay the counter overflow by a fraction of the reference period, αT_R. This can be done by gating the counter overflow with one pulse from a special counter that generates a wave of period, αT_R. Similar schemes can be constructed in several different ways; all require a clock frequency on the order of $1/\alpha T_R$. In this scheme, $T_R = \alpha T_R + N/f_v$, gives the lock condition. Hence, $\alpha = (f_v - Nf_R)/f_v$. For example, if a 100-Hz resolution is required at 400 MHz (f_v = 400.0001 MHz) then for an f_R of 100 kHz, $\alpha \simeq (1/4) \times 10^{-6}$, and hence, the required clock frequency is about 40 GHz, an impossible requirement. On the other hand, if maximum clock frequencies of 200 MHz are assumed permissible, then with 100-Hz resolution, the maximum VCO frequency can be 0.2 MHz. All schemes of this type suffer from the same restrictions on maximum usable VCO frequency, but the Digiphase scheme overcomes this restriction.

The principle is best illustrated by an example. Assume the required VCO frequency is 400.025 MHz, and the reference frequency is 100 kHz. The system is locked with N = 4000, then the initial VCO frequency is 400 MHz. If now N is changed to 4001 every fourth period (and reset to 4000 the rest of the time), initially, register A (Fig. 4), will take a little longer to overflow every fourth period. This produces a step change in voltage out of the phase detector every fourth period. Assuming V represents the voltage change out of the phase detector caused by changing N by unity, voltage into the VCO can be modeled initially as a DC level necessary to maintain 400 MHz together with a superimposed wave consisting of a quarter-period pulse of amplitude V. The DC component of this wave is exactly equal to 1/4 V. Generally, the DC component can be found by expansion of the resultant voltage wave at the VCO into a Fourier series. This will be found to be (T_R/T_t) V, where T_t is the period of incrementing N by unity (in this instance, $T_t = 4T_R$). Thus the average value of voltage applied to the VCO is that required to maintain 400 MHz plus 1/4 V, and since every V represents one count, i.e., 100 kHz in frequency, the voltage is exactly right to give 400.025 MHz out of the VCO. Hence, changing N to N + 1 once every M reference periods, and maintaining the value N the rest of the time, is equivalent to an average value of (N + 1/M).

Unfortunately, this procedure has an undesirable side effect in that it causes modulation sidebands to appear. This can be best explained by the preceding example. When steady state conditions are reached, f_v = 400.025 MHz, so when N = 4000, overflow is shorter than 10 μsec. During the first of the four periods, overflow is one-fourth of one clock count shorter, i.e., $(1/4) \times 2.5$ μsec = 0.625 μsec shorter. During the second period the overflow is again 0.625 μsec shorter, making a total lag of 1.25 μsec, and similarly, in the third period, the total lag is 1.875 μsec. During the last period the total lag accumulates to 2.5 μsec, i.e., one complete clock count. But during this period, N is increased by 1 making it necessary to count one

73

more before overflow. So when overflow does occur, the total lag has been effectively removed, making the overflow exactly equal to 10 μsec once every four periods. Whenever the period of overflow is not equal to 10 μsec the phase detector output registers a change in voltage. So under steady state conditions, a modulating staircase at a frequency equal to the fine subdivision of 100 kHz appears on the VCO.

In the example, modulation sidebands at 25 kHz and its harmonics will be present. The sidebands cannot be filtered out because of their close proximity to the carrier, but they can be eliminated in another way. Since the modulation is produced by a deterministic staircase ramp out of the phase detector, feeding a staircase ramp equal in magnitude, but opposite in phase to the summing point of the amplifier (Fig. 4) should, in principle, cancel the modulation.

The Digiphase synthesizer uses such cancellation. An external command, a binary equivalent of the desired frequency, is split into two parts: one representing a resolution to the nearest 100 kHz is fed into the B_r (B rough) register; the remainder into B_f (B fine). The purpose of B_r is to translate its input into presetting A, so that A divides by the proper integer, N. This is 4000 in the preceding example. B_f is an adder-accumulator. Its capacity for a VCO-fine resolution of 100 Hz and f_R = 100 kHz is 1000. The fractional portion (below 100 kHz) of the desired VCO frequency is added and accumulated in B_f once every reference period. After a sufficient number of reference periods, B_f overflows into the least significant bit of B_r. When this happens, the next presetting of B_r is one greater than the last presetting. B_f and B_r contain the following sequences in this example:

Number	B_f	B_r	Comments
1	250	4000	Conventional
2	500	4000	Conventional
3	750	4000	Conventional
4	0	4001	Overflow of B_f into LSB of B_r
5	250	4000	Conventional

B_f contains a staircase sequence of equally spread numbers: 0, 250, 500, 750 over the four periods, which is characteristic of any B_f. If the content of B_f is converted into an analog voltage by a D/A once every reference period, an analog staircase voltage is produced. This staircase, properly proportioned in magnitude, can be made equal to the modulation staircase out of the phase detector. Thus, with proper phasing, the D/A output, if added at the summing point, can cancel the phase detector modulation. The difficulty with this scheme lies in the high accuracy requirements in producing the canceling ramp. Also, any phase noise on the phase detector output offsets the canceling balance. It was found in practice that a sideband cancellation in excess of 40 dB was extremely difficult and unreliable.

2. Direct Synthesizer Combination

The PLL in the direct synthesizer combination (Fig. 5) is essentially the same as that shown in Fig. 3 except that a mixer stage and lowpass filter are inserted between the VCO and variable-capacity counter. The basic resolution of the loop is still f_R. A second synthesizer with an output from 0 to f_R in possible minimum steps of Δf is controlled by B_f. The two quadrature outputs of the synthesizer are combined in an SSB upconverter with a pure and stable signal of frequency, f_c. The output of the converter (frequency $f_c + f_s$) is combined in the mixer stage with the VCO output. Thus the signal entering A is at frequency $f_v - (f_c + f_s)$ and that coming out of A, at frequency $1/N\,[f_v - (f_c + f_s)]$. Since this frequency has to be equal to the reference f_R frequency when the loop is locked and in steady state, it follows that

$$f_v = Nf_R + f_c + f_s \quad .\tag{2}$$

In principle, f_c is not necessary since f_s has the required fine resolution and Nf_R provides the f_R steps. But if f_v is combined with f_s directly in an SSB system, the inevitable harmonics of f_s (due to system imperfections) appear around f_v and filtering of these requires, in effect, a Digiphase system. This problem is eliminated by

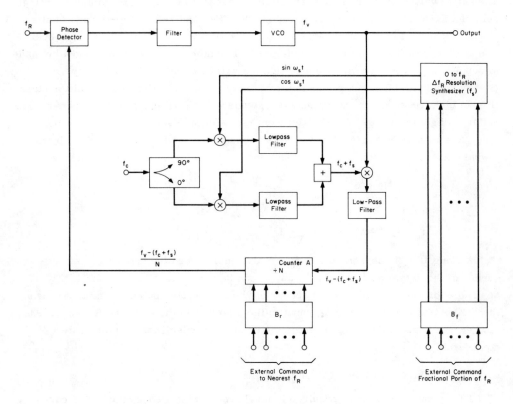

Fig. 5. Combination of two synthesizers to achieve high resolution.

offsetting f_s by f_c. The mixer sidebands are then at multiples of $f_c + f_s$. Also, with the present scheme, the SSB mixer has to cope with the fixed frequency, f_c, making the 90° coupler design much simpler. The lowpass filter has to pass $f_v - (f_c + f_s)$ and stop $f_v + f_c + f_s$. So to make the filter a relatively simple design, f_c should be an appreciable percentage of f_v.

Although, in principle, the scheme is capable of providing very fine resolution without the attendant drawback of close-in sidebands, care must be taken to avoid pickup that may cause such sidebands to appear. Care must also be taken to prevent $mf_v - nf_c$ (for low integer values of m and n) fall into the VCO frequency range. Omission of this precaution produces significant intermodulation sidebands. This becomes a severe problem when the percentage bandwidth of the VCO becomes large. However, the possible high frequency resolution, and the fact that it is independent of the main loop, makes this solution attractive for some applications.

3. High Frequency Extensions

The maximum operating frequency of the basic loop in the digitally controlled PLL (Fig. 4) is limited by the maximum working frequency of digital circuitry in the loop. The actual values here are being constantly revised upwards. At the time of writing, reliable divide-by-twos can be achieved at 600 MHz, and 1-GHz flip-flops (F/Fs) are about to become commercially available. But whatever this limit might be, if operation in excess of it is required, the simple PLL synthesizer discussed cannot be made to supply it. However, a simple modification will circumvent this difficulty.

By adding to the loop a power splitter, mixer, lowpass filter and amplifier stage (Fig. 6) high frequency operation may be extended. The RF input to the mixer is a signal at a single frequency, f_c, derived from the reference source. The input to divider A is thus at frequency $f_v - f_c$. It is assumed that $f_v + f_c$ is filtered out by the lowpass filter. As a consequence, under locked conditions, the VCO frequency is

$$f_v = Nf_R + f_c \qquad \text{if } f_v > f_c \tag{3}$$

and

$$f_v = f_c - Nf_R \qquad \text{if } f_v < f_c \quad . \tag{4}$$

As long as f_c is chosen so that $f_v - f_c$ does not exceed the upper working frequency of the digital circuitry in A, any f_v can be used. It is, of course, highly undesirable for any f_c component to appear on the synthesizer output. For this reason, the VCO output is used as the LO input to the mixer, and the f_c as the RF input. Also, the reverse transmissions through the mixer and power splitter will attenuate f_c considerably.

C. FINAL SYNTHESIZER SPECIFICATIONS

The synthesizer band is required to be above 1 GHz. Hence, the high frequency extension (Fig. 6) has to be used. The settling time, defined as the time it takes to transfer 90 percent of the energy in the synthesizer output signal to a new frequency

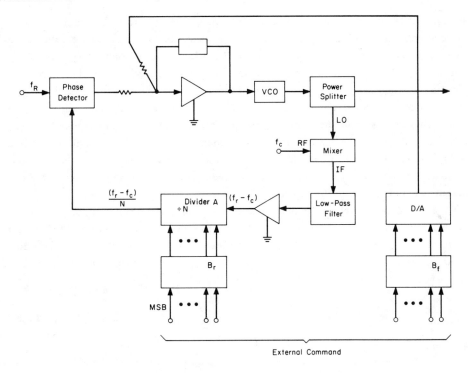

Fig. 6. **High frequency extension of digitally controlled PLL.**

after an external command is received to change frequency, must be 500 μsec or preferably less.

Assuming a synthesizer bandwidth of B Hz and the finest resolution of Δ Hz, the external command is a binary number of maximum increment value, B/Δ. For the lowest frequency, the all-0's state is used. All other possible states up to and including the all-1's state, are permissible and each one corresponds to a proportional frequency with all 1's corresponding to the highest available frequency. Thus the synthesizer provides B/Δ separate frequencies, spread Δ Hz apart. Let m be the number of bits in B/Δ.

D. DESIGN DETAILS

1. Digital Circuitry for Digiphase System

Digital circuitry in the system (Fig. 4) consists of two separately identifiable parts: overflow counter A, and control logic B. Their combined function is to accept an external command, as specified in the previous section, and translate it in the B stages to an acceptable presetting for the A stage and an appropriate D/A input. The A stage counts to overflow and is preset by B_r for the next cycle.

a. Overflow Counter A

The major requirement of this circuitry is that it count to the highest possible frequency, which means the "maximum counting speed" obtainable at almost any cost, or design. Alternatively, power consumption should be held to a minimum if it does not interfere with counting speed. These two requirements are usually contradictory. However, at the expense of orthodoxy and a slight increase in complexity, counting speed has been increased and power consumption reduced beyond that achievable by conventional design.

The counter includes a nonpresettable ECL divide-by-2 input stage and a presettable ECL divide-by-8 and divide-by-$(N/16)$ stages. The last stage may be built in lower speed logic, and consequently, consumes less power. The divide-by-2 is a very high speed F/F and is used merely to double the maximum acceptable frequency. Its presence or absence must, of course, be taken into account in the rest of the system, but it may or may not be used.

Because of the nature of the problem, gates cannot be eliminated completely without increased delay from other sources, and so the object of the design is to obtain a presettable divide-by-8 counter that has, at most, one F/F and one gate delay. Numerous approaches have been tried, most of them unsuccessfully. One acceptable design is shown in Fig. 7.

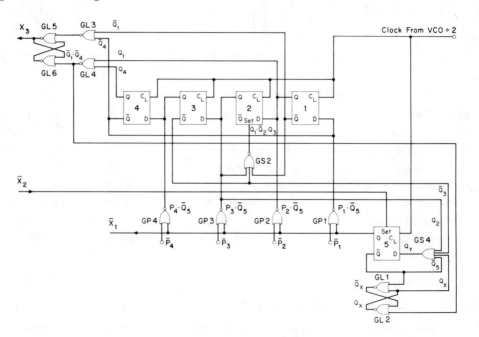

Fig. 7. High speed, presettable, ÷8 counter.

The counting operation is performed by a 4-bit shift register that does not require gating between stages as a conventional binary counter does. The shift register has two mutually exclusive counting sequences (Table 3). Once a sequence is initiated the register stays in that sequence as long as power is applied. However, the initial state of the F/F immediately after power is switched on is indeterminate, and so it is impossible to guarantee which sequence is available. Successful operation of the

Table 3. Shift Register Counting Sequence

Decimal Equivalent	Sequence I				Sequence II			
	Q_4	Q_3	Q_2	Q_1	Q_4	Q_3	Q_2	Q_1
0	0	0	0	0	0	0	1	0
1	0	0	0	1	0	1	0	1
2	0	0	1	1	1	0	1	1
3	0	1	1	1	0	1	1	0
4	1	1	1	1	1	1	0	1
5	1	1	1	0	1	0	1	0
6	1	1	0	0	0	1	0	0
7	1	0	0	0	1	0	0	1

system depends on the device always counting in sequence I. To assure this, a summing NOR gate, GS2, is connected to the set input on F/F 2. A high on the set input overrides all other inputs into the F/F and makes Q_2 go high. Because of GS2, this can happen only if the states of Q_1, Q_2, Q_3 are 1, 0, 1, respectively. This, in turn, can happen only during a sequence II count in decimal equivalent 1 or 4. When it does, the F/F set action makes Q_2 assume a 1 state putting the counter into decimal equivalent 3 or 4 of sequence I. So, other than the first few nanoseconds after power on, the counter counts in sequence I only.

The counter is required to count in its natural sequence from 0 to 7 without any presetting for most of its operating time (Fig. 7). However, when $\overline{X}_2$ (usually high) goes low, the counter is preset to some value determined by the $\overline{P}$ inputs ($\overline{P}_1$, $\overline{P}_2$, $\overline{P}_3$, $\overline{P}_4$) during the count sequence (Fig. 8) immediately following this event.

The $\overline{X}_2$ pulse comes from a slower portion of the system and is therefore shown with slow fall and rise times (Fig. 8). However, the transition times of $\overline{X}_2$ have to be less than six input clock pulses. The output of this stage, i.e., X_3 cannot, therefore, contain pulses of the same width as those driving the high speed stage. NOR gates, GL3 through GL6, ensure that X_3 is a 50-percent duty cycle wave of period equal to eight clock counts driving the high speed stage. This is only so during cycles not containing a preset. X_3 must, however, be acceptable to the low speed logic it drives, during all cycles including the preset one.

Presetting in the conventional manner implies forcing any two of the possible states to follow each other. If use was made of this method, it would be impossible to ensure that X_3 never contained a pulse as narrow as the input clock. So another presetting scheme was developed. The principle of the scheme (Fig. 9) is illustrated by the following example:

The counter's conventional sequence of operation is to go through all the states and to go to the 0000 state after the 1000 state. If, however, $\overline{X}_2$ goes low, the 1000 state becomes the preset enable state. The counter then goes to any state specified by the P inputs instead of to its natural

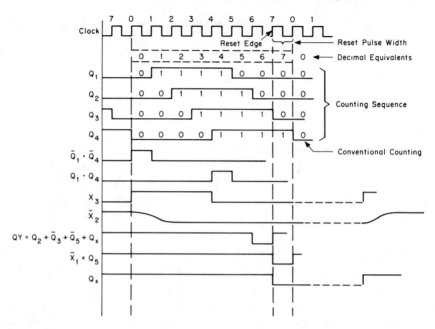

Fig. 8. Typical wave shapes for high speed ÷8 counter.

0000 state: For our design the 1110 state follows 1000. The counter then resumes conventional counting. The effect of this operation is to delay the 0000 state by a specified number of counts (three in the example presented). Since X_3 is produced by latch GL5 and GL6 (one of the inputs to which is the pulse $\overline{Q}_1 \cdot \overline{Q}_4$) X_3 will have an extended half period during the present sequence. This delay can vary from 1/8 to 7/8 of its conventional duty cycle. Thus X_3 can be guaranteed to never contain a pulse shorter than those generated during its conventional count sequence. The 0000 stage is used as the count state, i.e., the other input into the GL5/GL6 latch. In this way, all states of the counter can be preset. A presetting to the 0000 state is equivalent to having no presetting at all and letting the counter assume its natural state. Presetting is achieved through the four NOR gates: GP1 through GP4. The outputs of these gates are wired-OR into the D inputs of the F/F with the Q output of the preceding stage, and the first F/F with the $\overline{Q}$ output of the fourth stage. The four D inputs are thus:

$$D_1 = P_1 \cdot \overline{Q}_5 + \overline{Q}_4$$

$$D_2 = P_2 \cdot \overline{Q}_5 + Q_1$$

$$D_3 = P_3 \cdot \overline{Q}_5 + Q_2$$

$$D_4 = P_4 \cdot \overline{Q}_5 + Q_3 \quad . \tag{5}$$

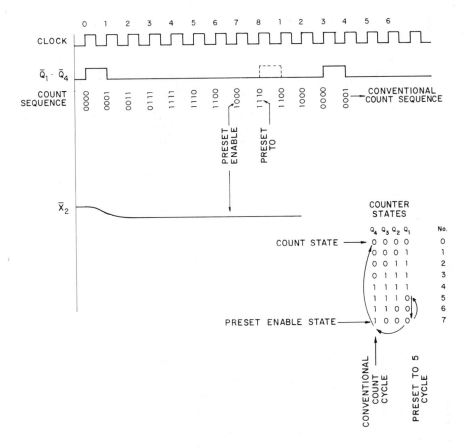

Fig. 9. Count sequence for preset to five.

Every time the 1000 state is reached, $D_i = P_i \cdot \overline{Q}_5$. During a conventional count, $\overline{Q}_5 = 0$ and the natural 0000 state follows. During the present cycle, $\overline{X}_2$ going low releases F/F 5 to respond to its D and clock inputs. During the sixth period of the preset cycle the D input on F/F (QY) goes low and rises again at the beginning of the seventh period (Fig. 8). Flip-flop 5 acts as a 1-bit shift register and so Q_5 goes low at the beginning of the seventh period and rises at the end of it. $\overline{Q}_5$ is one of the inputs to the latch composed of GL1 and GL2. The other input is the $\overline{Q}_1 \cdot \overline{Q}_4$ pulse, which goes high only during the 0000 state, thus QY, i.e., the D input to F/F 5 will not change state again until X_3 goes high again. The delay through the latch, GL5 and GL6, is offset by $\overline{Q}_5$ on GL1. The moment X_3 goes high, $\overline{X}_2$ follows suit, freezing F/F 5. The counter resumes its conventional sequence until $\overline{X}_2$ again goes low.

The delay in a particular presetting is one F/F delay, i.e., through F/Fs 1 to 4. At the same time, F/F 5 is also clocked, hence, the delay for the preset enable pulse (out of Q_5) is in parallel and does not add. Another delay is encountered in the

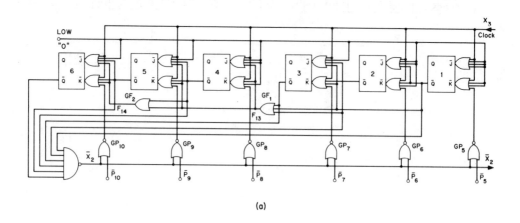

(a)

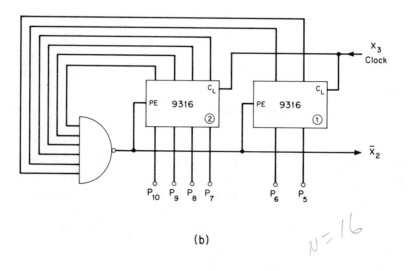

(b)

Fig. 10. Presettables (a) discrete $\div 2^6$ (b) LSI $\div 2^6$.

NOR gates, GP1 to TP4. Thus maximum setup delay is that through one F/F and one gate. Preset disable delay is only one F/F, i.e., F/F 5, which goes high on the next rising clock edge disabling gates GP1 to GP4.

Another system designed around the same performance parameters uses both count sequences, counting in one sequence during its normal cycle of operation and switching to the other sequence during its preset cycle to count to a number specified by the P_i. At the end of this cycle, it exits naturally into the first sequence. This operation occurs via gates analogous to GS2 (Fig. 7). The advantage of this scheme over the one adopted is that only four F/Fs are used: F/F 5 and the latch (GL1 and GL2) are eliminated. On the other hand, several more gates have to be used in the output stage and external to the counter because of the arrangement of the presetting P_i. The two systems are equally as fast and consume equal power.

Following the high speed divide-by-8 is a lower speed presettable divide-by-(N/16) stage. The clock for this part is the X_3 output of the high speed stage. The frequency of X_3 is 1/16 that of the mixer output, and hence, is assumed low enough for use with TTL logic. The length of this counter obviously depends on N. A typical 6-bit presettable divider built with discrete JK F/Fs are shown in Fig. 10(a). The OR gates on the $\overline{J}$ and $\overline{K}$ inputs are an integral part of the F/F. The law governing operation of these F/Fs is shown in Table 4 and the equation derived from it in Eq. (6).

Table 4. Law Governing F/F Operation

$\overline{J}$	$\overline{K}$	Q^{n+1}
0	0	Q^n
0	1	1
1	0	0
1	1	Q^n

$$Q^{n+1} = J\overline{Q}^n + \overline{K}Q^n \quad . \tag{6}$$

If F/Fs of this type are connected so that:

$$\overline{J}_1 = \overline{K}_1 = 0$$
$$\overline{J}_2 = \overline{K}_2 = \overline{Q}_1$$
$$\overline{J}_3 = \overline{K}_3 = \overline{Q}_1 + \overline{Q}_2$$

$$\vdots$$

$$\overline{J}_N = \overline{K}_N = \overline{Q}_1 + \overline{Q}_2 + \overline{Q}_3 + \ldots + \overline{Q}_{N-1}$$

and N-bit binary counter results. This principle was used to construct the divider (Fig. 10). The gates have four inputs on each $\overline{J}$ and $\overline{K}$. One input is the clock (X_3

from high speed counter); another is a preset line from the $\overline{K}$ parts. Because of cumulative interconnections, no more inputs are available after three stages. GF1 and GF2 have to be used to realize seven additional feed forward gates. The outputs of F_{13} and F_{14} are:

$$F_{13} = \overline{Q}_1 + \overline{Q}_2 + \overline{Q}_3$$

$$F_{14} = \overline{Q}_1 + \overline{Q}_2 + \overline{Q}_3 + \overline{Q}_4 \quad . \tag{8}$$

The unused inputs are all tied to a low, i.e., a 0. The counter, if left unimpeded, will assume all states between all 0's and all 1's in a binary fashion. When the all-1's state is reached, $\overline{X}_2$ goes low, and as a consequence, the output from the preset gates $GP_i: X_2 \cdot P_i$ are equal to P_i. Since these gates are tied to the $\overline{K}$ inputs, it follows from Eq. (6) that during this state, $Q^{n+1} = P_i$. Thus the next count starts from whatever number appears on the P_i.

Using a presettable 4-bit divider, LSI module 9316, considerable size reduction is achieved. The circuit in Fig. 10(b) performs the same function as the one in Fig. 10(a). Since two 9316 modules form an 8-bit presettable divider, the first two bits on module 1 are not used. The only difference between the two systems is that in Fig. 10(a), $\overline{P}_i$ have to be supplied, whereas in Fig. 10(b) the P_i are used directly.

The overflow counter A consists of a combination of the high and low speed parts. Lines X_3 and $\overline{X}_2$ in Figs. 7 and 10, respectively, are the same. This combination is, in effect, a fully synchronous counter composed of only minimum high-speed, high-power logic. It is faster than conventional equipment built entirely of high-power logic, and certainly, lower in power consumption. The one drawback is the complex pattern required for the P_i to achieve the counting scheme. However, translation from the external command can be achieved with low-power logic, adding only minimally to power consumption.

Figure 6 and the foregoing discussion show that the purpose of counter A is to divide incoming frequency, $f_v - f_c$, by N, so that $f_v - f_c = Nf_R$. Division is achieved by registering one output count per N input counts. The presetting P_i must be such that the counter possesses exactly N different states before overflow (i.e., an output) occurs. The beginning of a count cycle starts with $\overline{X}_2$ going low. Two distinct situations can occur: (1) the number N is such that there is no presetting on the three least significant bits in the high-speed counter, and (2) there is a presetting on these three bits. For the first case, there is an X_3 clocking edge into the low-speed counter immediately following $\overline{X}_2$. In the second instant, this clocking edge is missing (Fig. 9) so there are α or $\alpha + 1$ clocking edges, respectively, available for the low speed part of the counter depending on whether or not N is divisible by 8. Since, on the other hand, division must be maintained at the proper value independent of any possible pecularities in the counting scheme, it follows that if the presetting on the low speed counter is P_L when N is divisible by 8, it has to be $P_L + 1$ when N is not divisible by 8. Ensuring that this is so reduces the system to one where no counts are missing.

If a divide-by-2 is present in A and the presettable part of the divider is denoted by N_{Br}, then it follows that the output frequency from A (i.e., f_R under locked conditions) is

$$f_R = \frac{f_c - f_v}{2N_{Br}} \tag{9}$$

and hence, $f_v = f_c - 2N_{Br}f_R$. Increasing N_{Br} by unity and denoting the new f_v by f_{v+1}

$$f_{v+1} = f_c - 2(N_{Br} + 1) f_R \quad .$$

It follows that

$$f_{v+1} - f_v = 2f_R \quad .$$

So if a divide-by-2 is used, the smallest possible increment in the VCO frequency is equal to twice the reference frequency.

Usually, f_v is not such that $f_c - f_v$ is exactly divisible by $2N_{Br}$. Under such conditions, f_v can be represented as the sum of an integral part, f_{vI} (modulo $2f_R$) and a fractional part, f_{VF}. From Eq. (9)

$$N_{Br} = \frac{f_c - f_{vI}}{2f_R} \tag{11}$$

and the two extreme values of N_{Br}, i.e., N_{max} and N_{min} are

$$N_{max} = \frac{f_c - f_{vI\,min}}{2f_R}$$

$$N_{min} = \frac{f_c - f_{vI\,max}}{2f_R} \quad . \tag{12}$$

If the total capacity of counter A is n bits, then the required presetting, P_r, for all n bits is given by:

$$P_r = 2^n - N_{Br} \quad . \tag{13}$$

The low-speed counter has a capacity, 2^{n-3}. Hence, $N_L = (2^{n-3} - P_L)$. Similarly, the high-speed divide ratio will be $2^3 - P_H$, giving

$$N_{Br} = (2^{n-3} - P_L) * 2^3 + (2^3 - P_H) \tag{14}$$

where P_L is the integer part of P_r modulo 8 and P_H the remainder.

Alternatively, given a required $N_{Br} = N_L * 2^3 + N_H$,

$$P_L = 2^{n-3} - N_L$$

$$P_H = 2^3 - N_H \quad . \tag{15}$$

b. Control Logic − Stage B

The purpose of this stage is to accept an external command and translate it into an acceptable presetting for overflow counter A. (A relation between the required N and the individual presetting was described in the previous subsection.) From the foregoing subsection and Figs. 7 and 10, it is clear that the presettings, P_i, are needed only during the duration of the preset enable pulse, i.e., $\overline{X}_2$ for the low speed, and X_1 for the high speed portions. Since X_1 is contained within $\overline{X}_2$, the latter determines the necessary duration of the P_i. Typically, $\overline{X}_2$ is low for several hundred nanoseconds or less, and high for the remainder of the reference period (several μsec). Thus, a lot of time is available to set up an appropriate P_i from the incoming command, and consequently, low speed and very low power logic can be used.

The external command, denoted by M say, is a binary number (Sect. C). An all-0 M is assumed to correspond to the lowest VCO frequency, $f_{v\,min}$. The 1's state on M corresponds to the highest VCO frequency, $f_{v\,max}$, and all states in between are permissible. M can again be divided into two parts: one denoted by M_{Br} corresponding to the B_r part of the system, and the M_{Bf} corresponding to the B_f part. Consider for the moment only the M_{Br} part. The two extremes can be denoted by $M_{Br\,min}$ and $M_{Br\,max}$. The all-0 and all-1 settings, respectively. Thus

$$M_{Br\,min} \text{ implies } f_{vI\,min} \text{ implies } N_{max} \text{ implies } P_{r\,min} = 2^n - N_{max}$$

$$M_{Br\,max} \text{ implies } f_{vI\,max} \text{ implies } N_{min} \text{ implies } P_{r\,max} = 2^n - N_{min}$$

Since $M_{Br\,min}$ is zero, the foregoing is satisfied if

$$P_r = M_{Br} + (2^n - N_{max}) = M_{Br} + P_C \tag{16}$$

is satisfied. From Eqs. (12) and (16)

$$P_C = \left(2^n - \frac{f_c - f_{vI\,min}}{2 f_R} \right) \quad . \tag{17}$$

Since $M_{Br\,max}$ is the all-1's state, and Eq. (16) has to hold for all M_{Br}, P_r must contain at least one bit more than M_{Br} unless $P_C = 0$ (not usually the case). On the other hand, $P_C < 2 M_{Br\,max}$ since M_{Br} controls all B_r states. Thus $M_{Br\,max} < P_r < 2 M_{Br\,max}$ and so it follows that M_{Br} must be exactly one bit less than P_r, and since P_r contains n bits, M_{Br} will contain (n − 1) bits, and therefore, M_{Br} will contain (m − n + 1) bits.

M_{Bf} goes directly into an adder accumulator and from there through a D/A to the summing node of the first filter amplifier. Since this signal is to cancel a modulation produced by the B_f overflow, its phase must be reversed. This implies that

$$N_{Bf} = M_{Bf} \quad . \tag{18}$$

Equations (11) and (18) define all relevant relations between the input command and an applicable presetting. The B translate stages have to implement the foregoing considerations in logical hardware.

Since a number of separate steps are involved in the translation of M, a number of timing pulses (Fig. 11) used as clock pulses for individual translate steps are required. The total translation process has to take place during one reference period and the result must be available as a set of P_i during the X_1 and $\overline{X}_2$ pulses. So the setup process can start immediately after the $\overline{X}_2$ pulse goes high again. A convenient clock for the generation of the timing pulses is provided by X_3. The output of F/F A Or ed with X_2 provides a pulse that is high for the duration of X_2 and $\overline{Q}_A$. Pulse X_2 is obtained from $\overline{X}_2$ through an inverter. Since $\overline{Q}_A$ is obtained directly from $\overline{X}_2$, pulse $X_2 + \overline{Q}_A$ has no "glitch" in the middle. Also, $Q_B \cdot \overline{X}_2$ has a rising edge during the time that $X_2 + \overline{Q}_A$ is high. Q_B is $X_3/2$. The reason two possible $Q_B \cdot \overline{X}_2$ pulse trains are shown is that the phase of $X_3/2$ reverses during successive reference

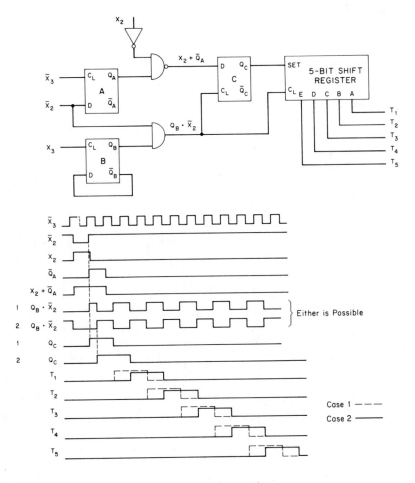

Fig. 11. Timing pulse generation.

periods, if X_3 contains an odd number of rising edges. This is equally as probable as containing an even number for an arbitrary external command. In principle, Q_B could be used as the clock for F/F C and the shift register. However, due to delays in the circuit, if the first situation existed, operation at the high frequency end is either intermittent or stopped. Combining Q_B with $\overline{X}_2$ shifts the first clock edge of Q_B by the width of $\overline{X}_2$ to ensure operation at all frequencies. The shift register's output is the five timing pulses (T_1 to T_5).

The period of X_3 is 16 times the period of the mixer output signal. $f_c - f_v$. The width of the timing pulses (Fig. 11) is twice the period of X_3, and therefore, 32 times that of $f_c - f_v$. Assuming the period is t sec wide, care must be taken to ensure that the reference period is larger than about 6t.

It is necessary to delay, by two reference periods, the B_f D/A output (Fig. 12). The reason for this is discussed in Sect. E1. M_{Bf} is entered through inverters into

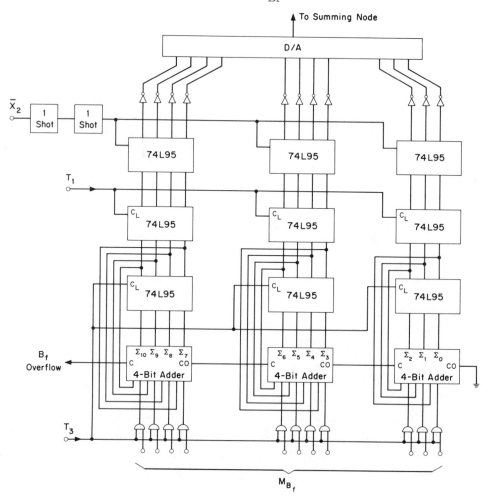

Fig. 12. B_f stage.

88

the adders. This number is entered into the first level of storage registers during timing pulse T_3 together with whatever number was contained in that level before. The contents of the first level is shifted info the second level by T_1, i.e., before the addition just described. Whatever was in the second level gets transferred into the third level by T_0, i.e., the output of the one-shot. Since the one-shot is controlled by the rising edge of $\overline{X}_2$, this transfer can happen before or after T_1, depending on the delay of T_0. This one-shot, therefore, provides a fine adjustment on timing of the transfer of B_f into the D/A, which can be slightly over or under two reference peri-ods, since progressively earlier pulses are used for each transfer. Low power logic can be used for this stage provided the timing pulses are wide enough. For this rea-son, T_3 and T_1 clock successive levels and not consecutive pulses such as T_2 and T_1.

The B_f overflow registered by the digital command for the overflow counter A presetting, occurs during T_3. This takes the form of a pulse that goes high and stays high for the duration of T_3 every time the adder overflows. If M_{Bf} is all 1's, this happens every reference period, but once every $2^{(m-n+1)}$ periods. On the other hand, no overflow occurs for an all-0 M_{Bf}. This means the VCO frequency is exactly divisible by $2f_R$. If M_{Bf} has 1 in its least significant bit and all 0's everywhere else, an overflow occurs every $2^{(m-n+1)}$ periods.

The B_r translate stage has to accept M_{Br}, any overflow from B_f, add them to-gether, and add the fixed offset P_C [Eq. (16)] to the sum. Next, it has to determine whether or not the resultant presetting is exactly divisible by 8, i.e., whether or not the three least significant bits are all 0. If they are, the sum on the adder output is the required value for further processing. If not, 8 has to be added to the existing sum. This is equivalent to adding 1 into the fourth bit. The structure of B_r differs considerably depending on the value of P_C and n. Therefore, Fig. 13 shows an 8-bit B_r stage with a $P_C = 60$. In this example, the B_r stage conforms with Figs. 7 and 10. Lines $\overline{P}_1$ through $\overline{P}_4$ coming out of the B_r stage are the same lines going into the high speed counter (Fig. 7). P_5 to P_{10} are connected directly to the preset parts of Fig. 10(b), or through inverters to the preset parts of Fig. 10(a).

The number, M_{Br}, entered on the B inputs of the 8-bit adder, is assumed to be always present. The number 60 (i.e., P_C) is entered on the A ports. A_2, A_3 and A_8 are hard wired to 0, 1, 0, respectively. F/F 1 is reset to zero by timing pulse T_1, i.e., $Q_1 = 0$ and hence, $\overline{Q}_1 = 1$. The T_1 pulse also sets $Q_B = 0$. So after T_1 happens, 00111100 (or 60 as a binary number) appears on the A_1 to A_8 inputs, is added to the already present M_{Br} and appears on the output as a sum (the Σ_i). During T_3, if there is a B_f overflow, the latch B flips over making $Q_B = 1$. This is also added to the sum. If there is no overflow, Q_B does not change state and the sum remains unchanged. The output of NAND gate G_1 is A + B + C. So the D input of F/F 1 is high if any one of the three least significant bits (A, B or C) is high. This is the case if the number to be preset is not divisible by 8. When D is high, timing pulse T_5 clocks a 1 into Q_1, and $\overline{Q}_1$ becomes 0. The only difference between 60 and 68 is in the four bits enclosed by dotted lines (Fig. 13). By flipping F/F 1 over, the change

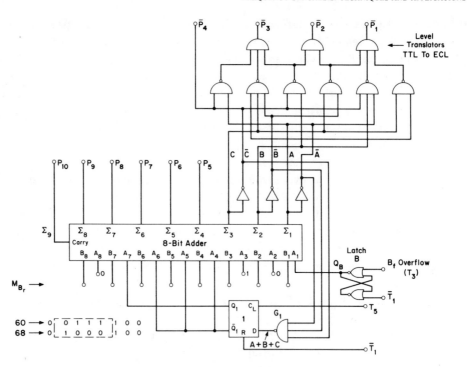

Fig. 13. B_r translate stage.

necessary to transform 60 into 68 is achieved. So during T_5, if any of the three least significant bits of the sum is a 1, 68 is added to the sum instead of 60.

Because of the shift register mode of counting in the high speed presettable divider, the three least significant bits in the 8-bit adder output have to be translated into the 4-line presetting, $\overline{P}_1$ to $\overline{P}_4$ (Table 5). The 3- to 4-line translator has to accept the contents of the three least significant bits on its input and produce the corresponding complement entry on its output. Forming maxterms gives:

$$\overline{P}_1 = (\overline{A} + C)(\overline{B} + C)(A + B + \overline{C}) \tag{19}$$

$$\overline{P}_2 = (\overline{B} + C)(B + \overline{C}) \tag{20}$$

$$P_3 = (B + \overline{C})(A + \overline{C})(\overline{A} + \overline{B} + C) \tag{21}$$

$$\overline{P}_4 = \overline{C} \quad . \tag{22}$$

The high speed part of divider A is built in ECL. The B_r adder, on the other hand, is in TTL. So, in addition to the 3- to 4-line translation, a TTL to ECL level change has to be effected as well. Level translators also perform the AND function. This feature was used to implement Eqs. (19) to (22). The top layer of AND gates in the translator are part of the realization of Eqs. (19) to (22) and change the TTL levels on their inputs to ECL levels on their outputs (Fig. 13).

Table 5. Translating Bits to Presettings

Three Least Significant Bits			Required Presettings				Complements			
C	B	A	P_4	P_3	P_2	P_1	P_4	P_3	P_2	P_1
0	0	0	0	0	0	0	1	1	1	1
0	0	1	0	0	0	1	1	1	1	0
0	1	0	0	0	1	1	1	1	0	0
0	1	1	0	1	1	1	1	0	0	0
1	0	0	1	1	1	1	0	0	0	0
1	0	1	1	1	1	0	0	0	0	1
1	1	0	1	1	0	0	0	0	1	1
1	1	1	1	0	0	0	0	1	1	1

E. TWO-SYNTHESIZER LOOP

The B_r command translator and presentable counter A (Fig. 5) are identical to those described in the preceding section. However, here, B_f does not overflow into B_r, and consequently, no staircase ramp is generated. The fractional portion (modulo f_R) of the input command goes into a command translator (B_f) which performs the same role for the secondary fine-resolution synthesizer as B_r does for the main loop. Assuming the fine frequency resolution required to be $\triangle$ Hz, the nearest acceptable $\triangle_o$ should be chosen, such that

$$\triangle_o = f_R/(2^k - 1)$$

where k is an integer. The external command into B_f will then be k bits long. The system can be arranged to produce the lowest required frequency out of the fine resolution synthesizer for an all-0 command and the highest frequency when all 1's appear on the B_f input. In the event that the lowest frequency required is 0 Hz, the B_f stage is just a straight-through connection.

Construction of the digital synthesizer is described next.

F. ALL-DIGITAL SYNTHESIZER

The basic principle used here is that of a pointer rotating at a fixed angular speed whose displacement-with-time plot is a sine wave. This process is implemented digitally by generating a sequence of numbers, appropriately spaced in time, that correspond to the vertical displacement of the pointer. These numbers, applied to the inputs of a D/A converter, give the actual samples of the required sine wave. A lowpass filter completes the process.

Simplicity of design constrains the number sequence generation to be done linearly, so direct implementation of the scheme requires ROMs that are used as look-up tables for the actual values of the sine function. In effect, such a scheme generates a triangular wave first (numerically) and then, via the ROM, translates it into a

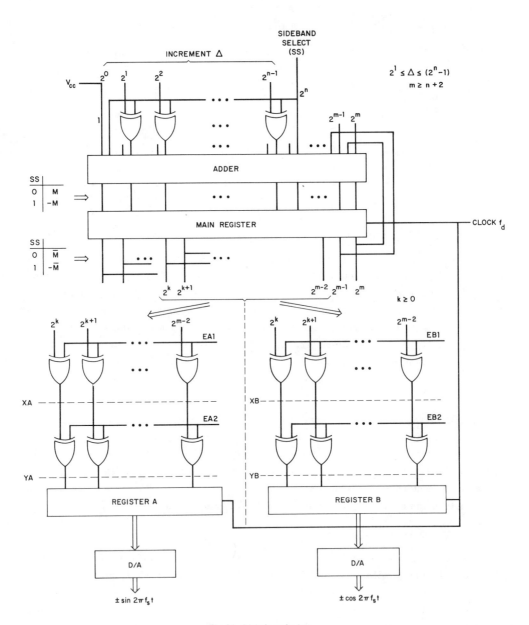

Fig. 14. Digital synthesizer.

sine wave. A perfect triangular wave has 81 percent of its energy concentrated in the fundamental component, and the nearest harmonic (third) has only 9 percent of the energy. In practice, some energy is present also in the 2nd harmonic. If the band of required frequencies is sufficiently narrow to exclude all harmonics with a reasonable margin, the ROMs can be dispensed with and direct filtering of the triangular wave will provide the required output. The synthesizer described here is of this type.

The output (Fig. 5) of the digital synthesizer (DS) is combined in a SSB system with the main loop frequency. So for a given fine resolution bandwidth, BF say, the DS needs to generate only $1/2$ BF if a sign reversal on the $\sin \omega_s t$ channel is available. For positive signs, output of the SSB is the upper sideband, $f_c + 1/2$ BF; and for negative signs, the lower one, $f_c - 1/2$ BF, giving a total coverage of the required BF Hz.

A further very effective scheme resulting in substantial reductions of the required BF is to construct the synthesizer in three levels: Basic loop, Digiphase, DS. The gist of the scheme is to use 1 or 2 (or more) bits of Digiphase modulation to resolve $1/2$ to $1/4$ (or more) times the basic loop reference frequency, f_R, but rely on the loop filter to suppress the unwanted sidebands. This is quite feasible since f_R is usually several hundred kilohertz, and the closed loop bandwidth at least an order of magnitude smaller, so even $1/2 \, f_R$ is well inside the loop stopband. So for a scheme with 2 bits of Digiphase modulation and sideband reversal, BF = $1/8 \, f_R$.

In a schematic diagram for the DS (Fig. 14), increment Δ is an n-bit number, the size of which determines the DS frequency output. Let δ be the fraction of the total register capacity corresponding to the LSB, then the smallest increment is 2δ. The two most significant bits, $m - 1$ and m, are used as quadrant control. Since, therefore, at least one sample per quadrant is used, the upper bound on the permissible output frequency, f_s, is half of the Nyquist rate, or equivalently, $1/2 \, f_D$ (where f_D is the DS clocking rate). It follows that

$$\delta = f_D * 2^{-(m+1)}$$

and since the smallest numerical increment is 2δ, the smallest frequency increment of the DS, δ_f say, is

$$\delta_f = f_D * 2^{-m} \quad .$$

Two cases of interest arise:

Case I where $m = n + 2$

This defines the maximal use of the DS giving a possible output frequency, f_s, in the range

$$0 \leqslant f_s \leqslant 1/4 \, f_D - \delta_f = \delta_f (2^{m-2} - 1) = \delta_f (2^n - 1) \quad .$$

Case II where $m = q + n + 1$; $q > 0$

$$f_L < f_s < f_L + \delta_f(2^n - 1)$$

and where f_L may be zero or $\delta_f * 2^n \leqslant f_L \leqslant \delta_f * 2^n(2^q - 1)$.

Usually, for case II, f_L is selected as some nonzero value (fixed or varied according to some required scheme). This makes it possible to have the DS supply a band of frequencies with the desired resolution, but with the lowest frequency not equal to zero.

The system is connected (Fig. 14) so that for upper sideband operation (Sideband Select, SS = 0) the adder and main register outputs are M and $\overline{M}$, respectively, and for lower sideband operation (SS = 1) these outputs are $-M$ and $-\overline{M}$. The negative sign represents 2's complement. As stated, the two most significant bits, $m - 1$ and m, are used for quadrant control, hence the MSB used for the two channels is $m - 2$. The number of bits that can be used is constrained by the size of the D/As. It may happen that the resolution, together with the required maximum frequency, results in $m - 2$ longer than the available capacity of the D/As. In such a case, $k > 0$ and only the most significant bits are used from 2^{m-2} down to 2^{m-2-r}, where r defines the D/A capacity. It turns out that this is not necessarily a disadvantage. The maximum resolution of the system is independent (within limits) of r. The effect of finite r is that sidebands appear due to truncation in the outputs. However, if a given sideband level is deemed acceptable (this defines the required r value) addition of finer resolution bits has no effect on these sidebands. In principle, therefore, for a given sideband level, arbitrary resolution is possible with finite D/A capacity. Table 6 shows the relations between the various enable lines, EA1, EA2, EB1, EB2, and the SS together with quadrant control bits, $m - 1$ and m. Also, the resultant X and Y words (defined in Fig. 14) are given.

Table 6. Relations of Enable Lines, SS and Quadrant Control Bits

SS	2^m	2^{m-1}	EA1	EA2	EB1	EB2	XA	YA	XB	YB
0	0	0	1	0	0	0	M	M	$\overline{M}$	$\overline{M}$
0	0	1	0	0	1	1	$\overline{M}$	$\overline{M}$	M	$\overline{M}$
0	1	0	1	1	0	1	M	$\overline{M}$	$\overline{M}$	M
0	1	1	0	1	1	0	$\overline{M}$	M	M	M
1	0	0	1	1	0	0	$-M$	$-\overline{M}$	$-\overline{M}$	$-\overline{M}$
1	0	1	0	1	1	1	$-\overline{M}$	$-M$	$-M$	$-\overline{M}$
1	1	0	1	0	0	1	$-M$	$-M$	$-\overline{M}$	$-M$
1	1	1	0	0	1	0	$-\overline{M}$	$-\overline{M}$	$-M$	$-M$

YA and YB are the words in register A and B, respectively. Since M is a word that increases in value by Δ with every clock-pulse, f_o, the four possible states of M appearing in Table 6 represent, respectively,

- M   positive, increasing word
- $\overline{M}$ positive, decreasing word
- $-M$ negative, increasing word
- $-\overline{M}$ negative, decreasing word

Based on the foregoing, together with the YA and YB events in Table 6, DS outputs are of the form shown in Fig. 15. For upper sideband operation (SS = 0) the solid lines hold. In reality, both channels are reversed in sign. Since the cosine is an even function of its argument, this has no effect on channel B (the cosine output), only channel A reverses in sign. This is indicated by the dotted lines.

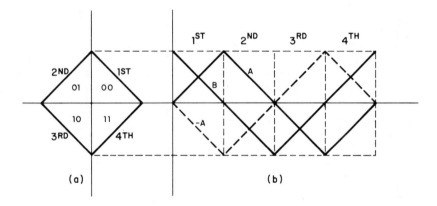

Fig. 15. Effect of quadrant control bits on outputs.

The circuitry to convert the first three columns of Table 6 into the four enable lines is shown in Fig. 16.

It is not necessary to have two channels (A and B) as shown in Fig. 14. One channel followed by a 90° phase splitter will provide the same results. However, analog switches are necessary for channel reversal (sine to cosine and vice versa). Depending on the application, this may not be a severe drawback. The advantage of the two-channel system is that an all-digital switching of the sine channel is effectively possible.

1. Analog Circuitry

The main components of the analog circuitry comprising the rest of the loop are: VCO, phase detector, loop filter, and pretuning and VCO linearizer circuitry. The basic loop equations can be derived as follows (Fig. 3):

$$f_v = k_v v_2 \tag{23}$$

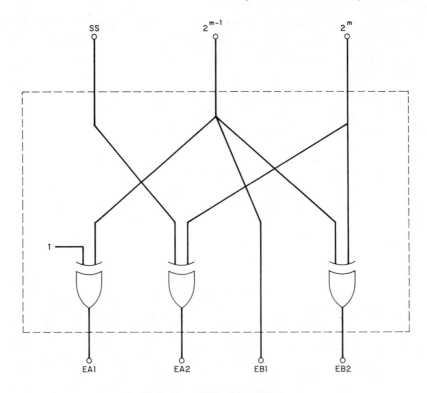

Fig. 16. Enable control for digital synthesizer.

hence

$$\phi_v = \frac{1}{s} k_v v_2 \quad . \tag{24}$$

The divider output frequency under locked conditions is $(f_v/N) \exp[-sT]$. The exponential part, $\exp[-sT_R]$, represents the inevitable one period delay through the divide by N. The phase detector produces an output voltage, v_1, that is proportional to the difference in the phases of the two inputs. The constant of proportionality is k_p, the phase detector constant, and

$$v_1 = k_p \left\{ \phi_R - \phi_v \frac{\exp[-sT_R]}{N} \right\} \quad . \tag{25}$$

The loop filter transfer function, $k_{DC} F(s)$, is defined by

$$v_2 = k_{DC} F(s) v_1 \tag{26}$$

and

$$F(0) = 1 \quad . \tag{27}$$

Combining Eqs. (23) through (26) and denoting

$$k_L = \frac{k_p k_{DC} k_v}{N} \tag{28}$$

the loop transfer function becomes

$$H(s) = \frac{\phi_v}{\phi_R} = \frac{N}{\frac{s}{k_L F(s)} + \exp[-sT_R]} = NQ(s) \tag{29}$$

then it follows from Eq. (27) that

$$H(o) = N \quad . \tag{30}$$

The dimension of the constants are:

k_p	Volt/radian	k_v	Radians/sec/volt
k_{DC}	Voltage ratio, hence dimensionless	N	Frequency phase ratio, hence dimensionless

From the foregoing it follows that

$$k_L \text{ has dimensions of } \frac{\text{volt} \times \text{radian}}{\text{radian} \times \text{sec} \times \text{volt}} = \frac{1}{\text{sec}} \quad . \tag{31}$$

Hence the loop constant, k_L, has dimensions of reciprocal time or equivalent frequency in Hz.

The choice of the loop filter, $F(s)$, determines the dynamic characteristics of the loop and affects the noise performance. An important parameter is the response of the system to a command to change frequency. Since a new command is an instantaneous change in N it can be modeled as a step change in phase. Ideally, this response should be as rapid as possible and the choice of $F(s)$ should reflect this requirement. It is well known that frequency selective networks have a delay in the step response proportional to the complexity of the network. Also any kind of frequency response shaping may result in a ringing step response thereby prolonging the effective settling time still further. At first glance, it would appear that the simplest kind of filter, or no filter at all, is desirable. This is undoubtedly true concerning settling time; however, there are other considerations that make it necessary to use a reasonably complex $F(s)$. At this stage, if no $F(s)$ is used, this does not mean the system will have an arbitrarily short settling time. Replacing $F(s)$ by 1 in Eq. (29):

$$H(s) = \frac{N}{\frac{1}{k_L} s + \exp[-sT_R]} \quad .$$

Frequency selectivity is provided by the delay, T_R and s/k_L. The system has a finite bandwidth, and hence, a nonzero settling time.

$F(s)$, of all the subsystems in the loop, is the most flexible since its hardware implementation is the least encumbered by realizability constraints. It should, therefore, not be selected until all other constituent parts have been determined optimally.

The phase detector determines the control voltage fed to the VCO, a very sensitive part of the system. Since the VCO phase is N times that of the reference signal, any noise out of the phase detector appears N-fold on the VCO output signal.

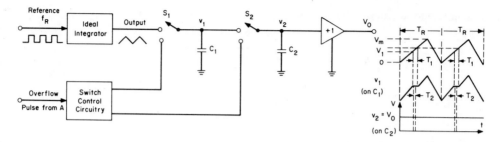

Fig. 17. Sample-and-hold phase detector.

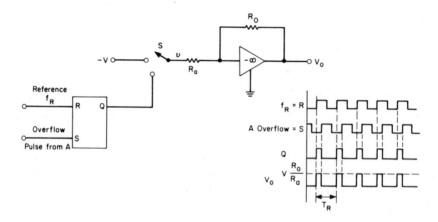

Fig. 18. R-S flip-flop phase detector.

Two different phase detectors were considered (Figs. 17 and 18). The sample-and-hold version is attractive on first sight. The reference signal, assumed to be square, is passed through an ideal integrator and transformed into a triangular wave. The overflow from counter A, via additional circuitry, controls switches S_1 and S_2. Switch S_1 is closed most of the time except at the instant of counter A overflow when it stays open for interval, T_1. This is short, a few percent compared with reference period, T_R. The excursion of the triangular wave is from 0 V to a maximum of V_m and reaches V_1 the instant S_1 closes. Therefore, voltage on C_1 resembles the form shown in Fig. 17. The second switch, open most of the time, is only closed for the interval T_2, where $T_2 < T_1$ and occurs during T_1. The voltage sampled by C_2 is therefore a constant, V_1.

If the system is in lock, there are no cycle-to-cycle variations and voltage, V_0, is the constant DC level necessary to maintain the required frequency. In principle, the first switch can be omitted. However, voltage V_0 then contains a slight modulation due to the rising nature of the sampled voltage. Re-introduction of S_1 decreases this modulation.

In theory, the system performs the phase detector function in that it provides a DC voltage output, proportional to the difference in the phases of its two input signals. If these signals have a phase difference that varies periodically, the amplitude of the phase detector output signal is proportional to the phase difference, and hence, varies with the same periodicity as the phase difference. In principle, no component at the referencing frequency is present in the output (especially, if two switches are used). To approach this ideal condition, as closely as possible, care must be taken that no resistive paths-to-ground exist across the capacitors. This is especially important with C_2. The amplifier should be an FET input device and have very small bias leakage current. Linearity of the triangular wave is also important. Departure from the linear case introduces a nonlinear relationship between output of the device and phases of the two inputs making the loop characteristics dependent on the operating point. This is very undesirable, especially for adequate B_f modulation cancellation. A very straightforward way to achieve an integrator is to use a very high gain amplifier bridged by a capacitor, C, and resistor, R, in series with the amplifier input. The output voltage of such an amplifier, $v_o(t)$, is related to its input, $v_i(t)$, by

$$v_o(t) = \frac{1}{RC} \int v_i(t) \, dt \quad . \tag{32}$$

The problem with this approach is that Eq. (32) holds only if amplifier gain is infinitely high. In practice, an adequate approximation can be obtained only at low frequencies, so this solution gives poor results for reference signals in the hundreds of kHz range. A passive integrator consisting of series resistor, R, and shunt capacity, C, does not suffer from any frequency limitation, but its linearity is poor unless the integration time, $t \ll RC$. It can be shown that when the output voltage reaches v_1, the linearity is proportional to the ratio (V_m/v_1) where V_m is the maximum voltage value that the output reaches. Thus for 1 percent linearity, $v_m = 0.01 \, V_m$. As only relatively small V_m (a few volts) are realistic, the voltage available from a phase detector using a passive integrator for high linearities is very small.

A bootstrap circuit is best for high quality linear ramps. A number of different versions of the circuit exist. Ultralinear ramp generators designed for TV could also be used for high quality performance. However, the complexity of these solutions could become prohibitive.

A much simpler phase detector (Fig. 18) consists of an edge-triggered R-S F/F whose output controls switch, S. One side of the switch is connected to a stable voltage source, $-V$, the other through resistor, R_a, connects to the summing point of a summing amplifier. The summing point is at ground potential, so when the switch is open, $v = 0$, and so is V_o. When the switch is closed ($v = -V$) then $V_o = (R_o/R_a) \, V$. If the R and S inputs are as shown in Fig. 18, i.e., equal in frequency, but shifted relative to each other in phase, the F/F output, i.e., Q assumes the shape shown: a rectangular wave of frequency equal to the two inputs, and the duty cycle is equal to the phase difference of the two incoming signals. If the two inputs are not at the same frequency, Q has a varying duty cycle at a rate equal to the difference in frequency of the two inputs. The instantaneous mark-space ratio over any one reference period is still equal to the phase difference in the two inputs. When $Q = 1$, switch S is closed, and when $Q = 0$, S is open. Therefore, V_o, on the time axis, is an exact replica of Q. The reason for $-V$ is to assure that the magnitude of the voltage on V_o is constant and noise free. The area under the V_o wave is directly proportional to the phase differ- ence between the reference signal and counter A overflow. A DC voltage proportional to this phase difference can be extracted from V_o by a lowpass filter. This phase de- tector realization followed by a lowpass filter is therefore equivalent to the sample- and-hold realization.

Practical experience indicates that for narrowband synthesizers with not very stringent B_f requirements, the sample-and-hold phase detector is a good choice. For wider bandwidth and large B_f the R-S solution is superior, despite more stringent loop filter requirements.

The VCO is a variable frequency oscillator tuned by a voltage. This voltage is applied to a varactor and the linearity of the frequency vs voltage curve is essentially the linearity of the varactor. For reasonably wideband VCOs, linearity is not very good. For the present context, the only meaningful definition of linearity is in terms of the slope of the VCO frequency-voltage curve. This follows from the fact that k_v, the VCO transfer constant, is the relevant parameter defining the VCO in the loop equations and k_v also happens to be the slope of the VCO frequency-voltage curve.

Any change in that slope over the frequency band of interest produces a directly proportional change in the loop constant, k_L [Eqs. (28-31)] and as a result an equal change in the closed-loop bandwidth. So a requirement of not more than say x percent variation in closed-loop bandwidth over the synthesizer band, is equivalent to an x per- cent limit on the linearity of the slope of the VCO frequency-voltage curve (assuming all other loop parameters are ideal). VCO manufacturers define VCO linearity as the ratio of the largest departure (in volts) from a straight line for any curve to the total range of tuning voltage required. Using this definition, a one or two percent linearity is readily achievable even over reasonably wide frequency bands.

The foregoing sounds nice, but is meaningless for PLL applications since it gives no indication by how much and in what fashion the slope of the VCO curve varies. Practical measurements indicate that several percent in the conventional definition may mean several hundred percent in slope linearity. Care has to be taken therefore, when specifying a VCO, to avoid misunderstandings.

Loop calculations assume k_v is constant over the whole band. In practice, as just discussed, this assumption is not true, and therefore, if equal quality perform- ance over the whole band is required, some sort of linearizer is needed. For large bandwidths, k_V will be a very large number since only limited voltages are available due to dynamic range restrictions of the preceding stage. VCO sensitivity to any dis- turbance on its input, and that includes unwanted noise as well as the required tuning voltage, is proportional to k_V. So every reasonable effort should be made to decrease k_V. Since the bandwidth is usually fixed by system specifications, only the tuning voltage range can be increased to decrease k_V. This too is limited by the voltage available from operational amplifiers.

For the Digiphase system the B_f D/A output is fed to the summing point of the summing amplifier where it is required to cancel the modulation produced by the B_f overflow into B_r. The effect of B_f overflow into B_r is to produce, during the overflow period (i.e., between any two consecutive overflows) a shortened overflow every reference period (top trace, Fig. 19). In the example considered the system is assumed to be locked to a frequency with a fine resolution equal to 1/5 that of the fundamental loop. The overflow pattern assumes the form shown in the second trace. The third trace, Δ, shows the amount and pattern of the difference between the second trace and a hypothetical second trace for a signal with a fine resolution equal to the fine resolution of the loop. The available canceling modulation, i.e., the B_f D/A output is of the same form, irrespective of the VCO frequency, and depends only on the fractional part in B_f.

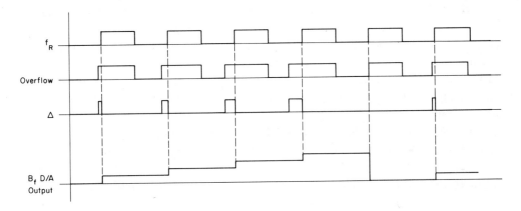

Fig. 19. B_f modulation.

Unfortunately, the same cannot be said about Δ. Although its periodicity depends only on the fractional part in B_f, the width of each pulse depends on the absolute value of $f_c - f_v$, since it can be shown that the first pulse in a sequence is equal in width to $g/(f_c - f_v)$; the second, to twice the width; the third, to three times the width, etc., where g is the fractional part in $f_c - f_v$ modulo f_R. In this example, $g = 1/5$. So if an appropriate $-V$ is chosen that cancels the modulation at any one frequency, cancellation is very poor at all other frequencies. It is fortunately possible to compensate exactly for this effect by making the voltage that is fed to switch, S (Fig. 18) proportional to N, i.e., the phase detector output wave magnitude varies with N, so although the pulses get wider as frequency, $f_c - f_v$, decreases, their magnitude decreases in the same proportion, making the area under each pulse frequency invariant. Since area under the pulse is proportional to the height of the sidebands produced by this modulation, a cancellation at one frequency holds at all others.

A further complication arises in that the two modulating voltages, although equal in period and strength (as described) are not of the same shape. Hence, a Fourier analysis of Δ and the B_f D/A output, yield series that have the same kind and number of components; however, corresponding components in the two series do not have the same magnitudes. This means that cancellation is effective over only some of the sidebands. The available degrees of freedom, i.e., the magnitude of $-V$ and the timing of occurrence of the B_f D/A output should be used up to adjust cancellation of the strongest sidebands. Fortunately, it can be shown that the closer the sidebands, the closer the two Fourier series match in magnitude. So for close-in sidebands, cancellation is fairly good even for high harmonics. As the sidebands spread apart, cancellation holds only within acceptable limits (say, 50 dB) for the first few. So in all cases, if an appropriate filter is used that attenuates frequencies where the cancellation is poor, good overall sideband suppression can be achieved. Fortunately, frequencies at which such a filter starts attenuating are not too close to cause unacceptable loop bandwidth. Detailed calculations are given in the next section.

The three layers of storage registers for the B_f digital subsystem cited in the previous section are necessary because: When an external command to change frequency is applied, this information only reaches the presetting during the first period, then a presetting is applied and it takes another period for the system to start reacting to this change in command. On the other hand, the output of the first B_f accumulator, if applied directly to the D/A, provides the first step of the modulation canceling ramp immediately. To offset the natural delay in the system then, the B_f output has to be delayed also. The maximum delay is about two loop periods. Its exact value is difficult to assess analytically because it depends on all inevitable strays in the loop. For this reason, the one-shot (Fig. 12) triggered by X_2 is used. This permits fine delay adjustment and is best done by monitoring the B_f sidebands on a spectrum analyzer. The delay is optimum when these sidebands are suppressed maximally. An analog system that incorporates these features is shown in Fig. 20.

A D/A converter controlled by M_{Br} (proportional to N) produces a voltage proportional to N with an offset, i.e., v_o can be denoted by

$$v_o = \delta_v (N - N_o) \tag{33}$$

where δ_v is the constant of proportionality and N_o represents the offset. The voltage, v', appearing on one side of the switch and v on the other are adjusted proportional to N by an appropriate choice of resistors, R_1, R_2, R_3 and R_a, as detailed in the next section.

The other path to the summing point goes via a set of resistors, R_5. The value of R_5 is, in effect, controlled by M_{Br} through the switches. R_5, together with R_4, which is connected to $-V$, provide open-loop pretuning. If R_5 is variable as well, a very close pretuning, taking into account the VCO nonlinearity, is possible. In this way, a voltage can be provided to the VCO that is very close to the actual voltage required to lock the loop. The loop itself supplies only a small difference voltage. Since the pretune is open loop, the voltage from it is applied almost at the same

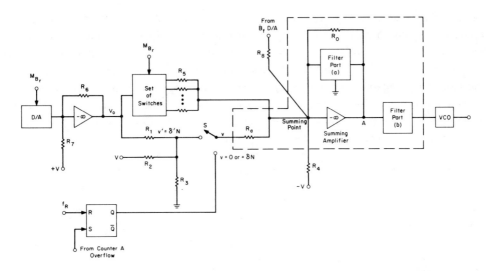

Fig. 20. Analog subsystem.

moment as the command to change frequency is given. So this scheme will increase the switching speed of the system.

The VCO constant, k_V, should be as linear as possible. This usually implies a linearizer in front of the VCO. A number of different schemes can be used. A straightforward one is shown in Fig. 20. A potential divider is formed with resistors R_{v1} and R_{v2}, with R_{v2} variable. The variation is controlled directly via switches from M_{Br}. A reasonably linear characteristic can be obtained, depending on the number of switches used.

An overriding concern in designing the loop is the system's stability. Consider the loop cut open at the input to the VCO (any other point will do equally as well) and assume that voltage, V_1, is applied to the right of the cut. Assuming the voltage fed around the loop and appearing to the left of the cut is V_2, the system is stable if V_2 is less than V_1 by the time a frequency is reached at which V_2 reverses in sign. This is equivalent to saying that for stability

$$\left| \frac{V_2}{V_1} \right| < 1 \qquad \text{for Arg} \left| \frac{V_2}{V_1} \right| = \pi \quad . \tag{34}$$

Assuming a cut is made as described

$$\frac{V_2}{V_1} = -k_L \frac{F(s)}{s} \exp[-sT_R] \quad . \tag{35}$$

Combining Eqs. (34) and (35) it follows that the one-point of V_2/V_1 cannot lie in the right-half s-plane for stability, or what is equivalent, that the 0's of

$$D(s) = 1 + k_L \frac{F(s)}{s} \exp[-sT_R] \tag{36}$$

lie in the left-half plane for system stability. From Eq. (29) it can be noted that the closed-loop poles are the same as the 0's of D(s). This is as it should be, because the poles of the closed-loop response are the natural modes of the system and must lie in the left-half s-plane for stability. Equation (35) is useful, however, for determining the loop phase and amplitude-stability margins. The phase margin (PM) is defined as the angle between the argument of V_2/V_1 and 180° at the frequency at which $V_2/V_1 = 1$. The amplitude margin (AM) is defined as the ratio of unity to the value $|V_2/V_1|$ assumes when its phase shift reaches 180°. Thus

$$PM = Arg \left\{ -k_L \frac{F(s)}{s} \exp[sT_R] \right\} \Bigg|_{s=j\omega_A} \tag{37}$$

and

$$AM = \frac{1}{\left| k_L \frac{F(s)}{s} \exp[-sT_R] \right|} \Bigg|_{s=j\omega_B} \tag{38}$$

where ω_A and ω_B are the angular frequencies at which $|V_2/V_1|$ reaches unit and Arg (V_2/V_1) reaches 180°. Since the overriding consideration is to maintain the system free of oscillation under any conceivable circumstances, acceptable margins will depend on system complexity and quality of components used. For a given loop considered, a calculation should be made to determine maximum amplitude and phase deviations as a result of worst-case component inaccuracies. These results, at least doubled, should give a comfortable safety margin.

Another important consideration in loop design is noise performance: noise generated (1) outside the loop, and (2) within the loop. Noise generated outside the loop implies noise on the reference line, since this is the only external input affecting the loop directly. From Eq. (29) it follows that a disturbance, $\Delta\phi_R$, on the reference produces a corresponding phase disturbance, $\Delta\phi_V$, on the VCO where

$$\Delta\phi_V = \Delta\phi_R \frac{N}{\frac{s}{k_L F(s)} + \exp[-sT_R]} . \tag{39}$$

At DC, $\Delta\phi_V = \Delta\phi_V = N\Delta\phi_R$ and as a consequence, $\Delta f_v = N\Delta f_R$. These relations follow directly from Eq. (39). Let the per unit error $\Delta\phi_V/\Delta\phi_R$ be denoted by ϕ_e. The steady state effect on the loop due to some transient perturbation, $\Delta\phi_R$, is given by

$$\lim_{t\to\infty} \phi_e(t) = \lim_{s\to0} s\phi_e(s)$$

$$= \lim_{s\to0} \frac{sNk_L F(s)}{s + k_L F(s) \exp[-sT_R]} . \tag{40}$$

So if no steady state effects are to be encountered in the loop from transient errors in the reference, F(0) = constant. This imposes a further requirement on the loop filter.

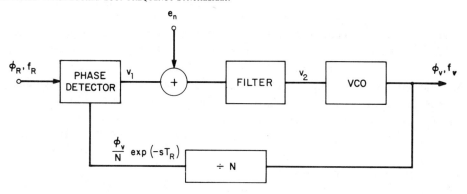

Fig. 21. Noise model.

Perturbations due to noise within the loop, whenever they occur, can be modeled as an equivalent additional noise voltage inserted immediately following the phase detector (Fig. 21). Though arbitrary, this model is convenient. If the effect of a disturbance anywhere else is analyzed, all that need be done is to refer it to an equivalent noise voltage, e_n.

The effect of e_n on the VCO phase is to produce a phase error, $\Delta\phi_V$. A simple analysis yields:

$$\frac{\Delta\phi_V}{e_n} = \frac{k_V k_{DC} F(s)}{s + k_L F(s) \exp[-sT_R]} \tag{41}$$

since $s\Delta\phi_V = \Delta\omega_V$, and where $\Delta\omega_V$ is the equivalent angular frequency error of the VCO, it follows that

$$\frac{\Delta\omega_V}{e_n} = \frac{sk_V k_{DC} F(s)}{s + k_L F(s) \exp[-sT_R]} = \frac{s}{k_p} H(s) \tag{42}$$

where $H(s)$ is the closed-loop response [Eq. (29)].

Assume the error voltage, e_n, has a component at a frequency, f_n, and let the deviation in f_V due to this component be denoted by Δf_{vn}. The index of modulation is

$$m = \frac{\Delta f_{vn}}{f_n} = \frac{\Delta\omega_{vn}}{\omega_n} \ . \tag{43}$$

For small m the sideband in the VCO output due to f_n at frequencies $f_V \pm f_n$ is given by

$$\text{sideband suppression} = 20 \log_{10} \left(\frac{m}{2}\right)\Big|_{f=f_n} \text{dB}$$

$$20 \log_{10} \left|\frac{H(j\omega_n) e_n}{2k_p}\right| \ . \tag{44}$$

For low frequencies (i.e., s very small) the error in the VCO frequency, $\Delta\omega_v$, can be seen to be very small [Eq. (42)]. Equation (47) states analytically what was said at the beginning of this chapter: the loop tracks out (eliminates) any DC errors

105

that occur in the loop and holds the VCO frequency locked onto the reference. Equation (42) further implies that this tracking out of errors is still very good at very low frequencies at which the magnitude of $sH(s)$ is still very small. This process deteriorates with increasing frequency and becomes a maximum where $F(s)$ is maximum. This happens, as will become apparent, at the closed-loop cutoff frequency. So the wider the loop bandwidth, the further out in frequency will the noise suppression reach. Beyond the cutoff point, loop attenuation will attenuate noise as well as other components. This leads to the further requirement on $F(s)$ is to provide sharp cutoff for the loop, which implies high-order filtering.

One interfering signal that merits attention and which is usually present in the loop if use is made of the Digiphase principle is the effect of B_f overflow. It was shown that if the fine resolution was f_R/m, then the counter would overflow ΔT earlier than the reference period, T_R, where

$$\Delta T = \frac{1}{m(f_c - f_v)} \quad .$$

The VCO frequency $(f_c - f_v)$ can be represented as having an integral portion exactly divisible by N and equal to Nf_R, and a fractional portion, f_R/m, thus

$$\Delta T = \frac{1}{m(Nf_R + f_R/m)} = \frac{T_R}{mN + 1} = \frac{T_R}{mN} \quad . \tag{45}$$

The approximation in Eq. (45) is usually very good since N will be a large number and $m \geqslant 2$. The phase detector output, v (Fig. 20) is either zero, then switch, S, is open or is equal to δN. The net effective voltage applied to the following stages is δN multiplied by the duty cycle. (More calculation details are given in the next section.) Since the effective duty cycle of the first overflow deviation is $\Delta T/T_R$, the voltage due to it is

$$e_{nl} = \delta N \frac{\Delta T}{T_R} = \frac{\delta}{m} \quad . \tag{46}$$

The second overflow shortening causes a doubling in the total shortening and the effective instantaneous error voltage increases to $2e_{nl}$. For any given m there are $(m-1)$ overflow shortenings. (At the mth time N is increased by one and no shortening occurs.) The peak voltage therefore, due to the B_f modulation, is

$$e_n = (\frac{m-1}{m})\delta \quad . \tag{47}$$

Also the phase detector constant, k_p, is shown (next section) to be given by

$$k_p = \frac{\delta N}{2\pi} \quad .$$

Using this relation with Eqs. (44) and (47) gives the fundamental sideband suppression for a sideband frequency, f_R/m.

$$\left.\begin{array}{l} \text{First} \\ \text{Sideband} \\ \text{Suppression} \\ \text{for} \end{array}\right\} \frac{f_R}{m} = 20 \log_{10} \left| (\frac{m-1}{m}) \pi \frac{H(j\omega_m)}{N} \right|$$

$$= 20 \log_{10} \left| (\frac{m-1}{m}) \pi Q(j\omega_m) \right| . \tag{48}$$

2. Loop Calculations

It is convenient to consider the analog system (Fig. 20) first without the filter (enclosed in dotted lines), and then the loop filter alone. The summing amplifier part of the filter acts also as a summing point for several lines. Filter part (a) is an open circuit at DC, and the summing action of the amplifier is determined by R_o and the appropriate resistors coming into the summing node.

3. Phase Detector and VCO Calculations

Let the frequency band from $f_c - f_{v\,min}$ to $f_c - f_{v\,max}$ be denoted by f_{max} to f_{min}, and the center frequency by f_o. The corresponding divider ratios, N, are N_{max}, N_{min} and N_o. The D/A controlled by M_{Br} provides a voltage, v_o, at the amplifier output. This voltage is linearly proportional to N and should be adjusted to give a maximum voltage swing without incurring some nonlinear effects due to amplifier imperfections. The largest swings can be obtained by having v_o go from a positive to a negative voltage (or vice versa) such that maximum magnitudes are reached at the two extremes. If these are made equal to each other, v_o is zero at the center frequency, f_o. If the foregoing are satisfied,

$$v_o = \delta_v(N - N_o) \tag{49}$$

where δ_v is a constant of proportionality and N represents the particular divide ratio. Resistors R_6 and R_7 have to be chosen so that Eq. (49) is satisfied for some specified δ_v. This will always be possible. The voltage, V, at the end of R_7 may be positive or negative.

Consider the path through switch, S, to the summing point. Voltage, v, will either be zero when the switch is open, or it will assume some finite value when the switch is closed. This value must be proportional to N. Choosing the constant of proportionality to be δ,

$$v = 0 \qquad \text{when S is open}$$
$$v = \delta N \qquad \text{when S is closed} . \tag{50}$$

From Fig. 20, it follows that when S is closed

$$v = N = \frac{G_1}{\Sigma} v_o + \frac{G_2}{\Sigma} V \tag{51}$$

where $\Sigma = G_1 + G_2 + G_3 + G_a$ and the G stands for $1/R$.

Denoting $A_1 = G_1/\Sigma$ and $A_2 = G_2/\Sigma$ it follows from Eqs. (51) and (49):

$$v = \delta N = A_1 \delta_v(N - N_o) + A_2 V . \tag{52}$$

If this equation is to be satisfied for all N

$$\delta = A_1 \delta_v \text{ and } A_1 \delta_v N_o = A_2 V, \text{ hence}$$

$$A_1 = \frac{\delta}{\delta_v}$$

$$A_2 = \frac{\delta N_o}{V} \quad . \tag{53}$$

A_1 and A_2 are thus specified when δ, δ_v, N_o and V are specified. There are four resistors and only two constraints. Two other constraints necessary to determine a unique set of resistors come from the loop-gain requirements and the maximum usable voltage on switch, S. The loop-gain requirement fixes k_{DC}, which is proportional to R_o/R_a. So R_a should be left here as a free parameter to be determined later. The voltage on S fixes v'_{max}. Since v'_{max} will occur at N_{max}, the ratio v'_{max}/v_{max} is, in effect, determined.

Solving the equations existing between the G_i and A_1 and A_2 and expressing the results in resistor ratios, gives:

$$\frac{R_1}{R_a} = \frac{1 - (A_1 + A_2)}{A_1\left(1 + \dfrac{R_a}{R_3}\right)} \tag{54}$$

$$\frac{R_2}{R_a} = \frac{1 - (A_1 + A_2)}{A_2\left(1 + \dfrac{R_a}{R_3}\right)} \quad . \tag{55}$$

When switch, S, is open, v' is given by

$$v' = \frac{G_1}{G_1 + G_2 + G_3} v_o + \frac{G_2}{G_1 + G_2 + G_3} V$$

which can be rewritten as

$$v' = \frac{A_1\left(1 + \dfrac{R_a}{R_3}\right)}{\dfrac{R_a}{R_3} + (A_1 + A_2)} v_o + \frac{A_2\left(1 + \dfrac{R_a}{R_3}\right)}{\dfrac{R_a}{R_3} + (A_1 + A_2)} V$$

$$= B_1 v_o + B_2 V$$

$$= B_1 \delta_v (N - N_o) + B_2 V$$

$$= B_1 \delta_v N + B_1\left(\frac{B_2}{B_1} V - \delta_v N_o\right)$$

$$= B_1 \delta_v N = \delta' N \tag{56}$$

where $\delta' = B_1 \delta_v$, since $V(B_2/B_1) = V(A_2/A_1) = \delta_v N_0$. So v' is proportional to N if v is proportional to N. Also, since $B_1 > A_1$, it follows from Eq. (53) that $(v'/v) > 1$ always.

From Eqs. (53) and (56)

$$\frac{v'}{v} = \frac{1 + \frac{R_a}{R_3}}{\frac{R_a}{R_3} + (A_1 + A_2)} \quad . \tag{57}$$

This holds for all values of v'/v, hence also for v'_{max}/v_{max}. Solving for R_3/R_a gives

$$\frac{R_3}{R_a} = \frac{\dfrac{v'_{max}}{v_{max}} - 1}{1 - (A_1 + A_2)\dfrac{v'_{max}}{v_{max}}} \tag{58}$$

so once the ratio v'_{max}/v_{max} is specified, Eqs. (54), (55) and (58) together with Eq. (53) give the values of R_1, R_2 and R_3 in terms of R_a.

The voltage, V, is usually a convenient supply voltage, e.g., 15 V. δ_v is specified by the maximum swing of v_0, N_{max} and N_{min}. The only parameter still to be chosen is δ. Obviously, all resistors used must be positive, which imposes limits on the permissible range of δ. Also pretuning requirements impose certain constraints on δ.

The phase detector part of the system consists of S and the R-S F/F. The phase detector constant, k_p, defines the voltage available on the device output, i.e., v per given phase difference on the R-S F/F inputs. The maximum phase difference can be 2π radians, at which instant the maximum voltage, i.e., $v = \delta N$, will be available; thus, for a phase difference of ϕ_Δ, the voltage available is $(\delta N/2\pi)\phi_\Delta$. Hence the phase detector constant is given by:

$$k_p = \frac{\delta N}{2\pi} \quad . \tag{59}$$

Pretuning is accomplished by resistive path R_5 (Fig. 20) with switches controlled again by M_{Br}. Assume, momentarily, that only a single R_5 resistor is connected between v_0 and the summing point. The shape of the pretune characteristic, referred to the output of the summing amplifier, is fixed by the following equations:

$$V_{p\,max} = -v_{o\,max}\frac{R_o}{R_5} + V\frac{R_o}{R_4} \tag{60}$$

$$V_{p\,min} = -v_{o\,min}\frac{R_o}{R_5} + V\frac{R_o}{R_4} \tag{61}$$

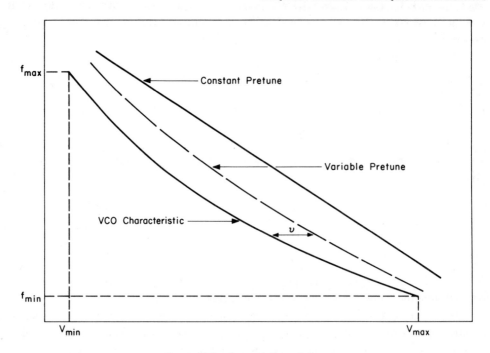

Fig. 22. VCO and pretune characteristics.

where v_p represents the voltages at point A (Fig. 20). These equations determine the straight-line pretune (Fig. 22). Solving for the two unknowns, R_4 and R_5, gives:

$$\frac{R_4}{R_o} = V \frac{(v_{o\,min} - v_{o\,max})}{(V_{p\,max}\,v_{o\,min} - V_{p\,min}\,v_{o\,max})} \tag{62}$$

$$\frac{R_5}{R_o} = \frac{(v_{o\,min} - v_{o\,max})}{(V_{p\,max} - V_{p\,min})} \quad . \tag{63}$$

Since only positive solutions are acceptable, the system has to be arranged so that the resulting v_o and V_p voltages always lead to positive R_4 and R_5. R_o is, of course, also a positive resistor. This will always be possible.

The horizontal gap between the straight-line pretune and the VCO characteristic (Fig. 22) represents voltage the phase detector must supply to achieve lock. If the VCO characteristic is not linear, as is almost always the case, the voltage required from the phase detector is not constant nor does it vary linearly with frequency. The maximum voltage from the phase detector is δN, which requires a closed switch. A slight perturbation in the system requiring more voltage from v would send the loop out of lock. The same is true of the other extreme. So a 50-percent duty cycle is best as a maximum safety margin. This implies that v would supply only $1/2\ \delta N$ volts. Obviously, for a nonlinear VCO characteristic, this can only be achieved at one point if the pretune is linear. On the other hand, a nonlinear pretune that follows the VCO characteristic so that the horizontal distance between it and the VCO characteristic varies linearly with frequency, can always be arranged to require only 1/2 δN out of the phase detector for all N. A realistic solution lies between the extremes of a straight line and an exactly fitting pretune characteristic.

Nonlinear pretuning can be accomplished in several ways. A conceptually straightforward way is to construct a nonlinear D/A that simply supplies the required voltage to the summing point. The drawbacks to this scheme are its complex construction and acceptability for only one VCO. If the VCO had to be replaced, a new D/A would have to be constructed. This need not be a problem, but a more flexible approach is preferable (Fig. 20).

Assume that a set of six switches controlled by logic levels is used. Taking the six most significant bits of M_{Br} as the switch drive, each bit is connected to one of the switches. A seventh resistor is connected across the whole system from v_o to δP. The loop is switched on and the duty cycle of v is adjusted by the seventh resistor until it is 50 percent and the resistor is at its maximum at the lowest frequency (i.e., when all the bits of M_{Br} are 0 and none of the other resistors are connected) or at a minimum at the highest frequency (when all bits of M_{Br} are 1 and all other resistors are connected). This requires some experimentation and adjustment by resistor, R_4. Then the frequency is changed progressively (by changing the external command) and the remaining six resistors adjusted so that the duty cycle of v stays as close to 50 percent as possible. This is done with the resistors in the system. The closeness with which one can approach the 50-percent duty cycle over the whole range is a function of VCO characteristic nonlinearity and the number of switches used in the pretune. In most cases, however, with six switches, 50-percent duty cycle within a few percent can be achieved.

The procedure just described for the calculation of the resistors in Fig. 20 is complicated by the Digiphase requirements. If a non-Digiphase system is used, similar to that shown in Fig. 5 for example, it is no longer necessary to maintain a voltage v (Fig. 20) that is proportional to N. The calculations become much simpler. The resistor R_1 can be dispensed with altogether and the switch is then supplied only by V through the potential divider formed by resistors R_2 and R_3. The magnitudes of the required voltages v' and v are again as dictated before by dynamic range and loop requirements. The M_{Br} D/A supplies only the pretune path through R_5. In a non-Digiphase system R_8 is of course not used.

4. Loop-Filter Calculations

Different loop-filter requirements appear throughout the previous discussion. If all are met, the system would be close to ideal. Unfortunately, it would also be unrealizable since some requirements are contradictory.

Filter Requirements

a. Because of dynamic range and manufacturer imposed VCO constraints the loop gain, k_L, can only lie between a very limited range, which imposes a severe constraint on the ratio of two of the loop-filter coefficients.

b.   No steady state errors from transient perturbations in the reference are permitted in the loop, hence $F(0)$ = constant.

c.   Short as possible settling time requires low order $F(s)$.

d.   Loop noise performance requires high order $F(s)$ with sharp selectivity.

e.   High order $F(s)$ must have a non-ringing step response.

f.   Stable, closed-loop system with reasonably high stability margins.

g.   Must work into a resistive divider shunted by some capacitance.

h.   Simple design. Unnecessary complexity introduces excessive noise and volume.

i.   Suppressed reference frequency and its harmonics in VCO output to 70 dB below the desired output. This implies a 120-dB suppression of the fundamental frequency and progressively less for the other harmonics because of loop gain.

Two fundamentally different compromises can be used. One, probably the more common, is to use a low first-order filter resulting in a second-order loop cutting off a decade or more below the reference frequency. Such a filter could be made to satisfy most of the requirements fairly well except requirements d and i as well as g. The high rejection required at f_R and its harmonics could be provided by additional traps (tuned circuits) at f_R and some of its higher harmonics, which is an inefficient and sensitive way to achieve rejection. A somewhat better way would be to build a second lowpass filter with only one attenuation pole at f_R. This results in a tradeoff between requirements h and i.

The second approach should achieve a better optimum. It tackles the problem by equating the closed-loop transfer function of the system $Q(s)$ to some desirable filtering function. This is the most rational approach since the poles of $Q(s)$ are the natural modes of interest and once these are well behaved no closed-loop stability problems will be encountered. It should be noted that the order of $Q(s)$ can be arbitrary as long as all poles lie in the left-half plane. Once $Q(s)$ is determined optimally from the loop-performance requirements $F(s)$, the filter-transfer function, will automatically satisfy optimally all filter requirements and its form is easily derivable from Eq. (29). $F(s)$ is then realized as a frequency-selective network and although it will not be a filter in the conventional sense, it is exactly the right function such that when the loop is closed the resulting closed-loop transfer function has the desired shape and performance. The procedure is best illustrated by an example. Let the closed-loop transfer function be of 6th order. Then

$$Q(s) = \frac{B_o}{s^6 + B_5 s^5 + B_4 s^4 + B_3 s^3 + B_2 s^2 + B_1 s + B_o} \; . \qquad (64)$$

Its step response $h(t)$ must settle within its final value as quickly as possible. Minimal ringing is therefore permitted. A gaussian response guarantees no ringing, but also has a shallow attenuation curve. However, it eventually reaches a rate of rejection of the order of 20 dB/decade. So the proposed filter should be capable of 120-dB rejection not too far beyond a decade of cutoff. Thus, if an appropriately low cutoff is used, requirements d, e and i can be satisfied. The form of Q(s) satisfies requirement b. If the other requirements permit Q(s) to be gaussian, requirement c is also met. The settling time of a gaussian filter of order n to a step response is approximately equal to $n/4f_c$ where f_c is the filter-cutoff frequency. This compares favorably with most other filters of the same order. For example, if f_c = 10 kHz, then for n = 6, the settling time is, according to the formula: 150 μsec.

Gaussian filters have quadratic factors with very low Q values, so they will lead to very insensitive networks to satisfy requirement h. Requirements h and i are contradictory so the lowest order that meets i constitutes an optimum compromise. A 6th-order function is such a compromise. The stability of the system is guaranteed by constraining the closed-loops natural modes to be in the left-half plane. The natural modes of the system are the poles of Q(s). So if Q(s) is gaussian, these will naturally lie in the left-half plane and because of the low Q-factors involved, stability margins will be fairly large when compared with the other approach.

Requirement g can usually be met. If the same is true of requirement a, a good compromise is achieved.

Let the function F(s) assume the form:

$$F(s) = \frac{A_o}{s^5 + A_4 s^4 + A_3 s^3 + A_2 s^2 + A_1 s + A_o} \qquad (65)$$

Then from Eq. (29) it follows that

$$Q(s) = \frac{1}{\dfrac{s(s^5 + A_4 s^4 + A_3 s^3 + A_2 s^2 + A_1 s + A_o)}{k_L A_o} + \exp[-sT_R]} \qquad (66)$$

Expanding $\exp[-sT_R] = 1 - sT_R + (1/2)s^2 T_R^2 \ldots$ and substituting in Eq. (66) gives:

$$Q(s) = \frac{k_L A_o}{s^6 + \left(A_4 - \dfrac{T_R^5}{120}\right)s^5 + \left(A_3 + \dfrac{T_R^4}{24}\right)s^4 + \left(A_2 - \dfrac{T_R^3}{6}\right)s^3 + \left(A_1 - \dfrac{T_R^2}{2}\right)s^2 + (A_o - T_R) + k_L A_o} \qquad (67)$$

Only up to the 5th-order term in the infinite series for $\exp[-sT_R]$ was used. This is a good approximation. T_R has to be normalized to the loop-filter cutoff, making T_R much less than unity.

Equations (64) and (67) are equivalent; hence, a one-to-one correspondence between the coefficients of F(s) and Q(s) exists. However, it will be seen from Eq. (67) that

$$B_o = k_L A_o$$

and

$$B_1 = A_o - T_R \quad .$$

Both k_L and T_R are fixed by the loop requirements, so the ratio B_o/B_1 is also fixed.

$$\frac{B_o}{B_1} = \frac{k_L A_o}{A_o - T_R} \simeq k_L \quad . \tag{68}$$

This expression is requirement a expressed in analytical form. If Eq. (68) holds, or at least leads to an acceptable value of k_L, the problem is solved; if not, a function that does satisfy Eq. (68) has to be found. There is no analytical procedure that can be followed to give a function of the type required. The only possible approach is to use an iterative optimization procedure via a computer, the most powerful method of which is described by Fletcher and Powell.[3]

Assume that an appropriate Q(s) as given in Eq. (64) has been found. The function F(s) is then readily found with the aid of Eq. (67). There are an infinite number of realizations, one of which (Fig. 23) contains two amplifiers, and its voltage transfer ratio can be expressed as the product of three parts, thus

$$\frac{V_2}{V_1} = T_1(s)\, T_2(s)\, T_3(s) \quad . \tag{69}$$

The transfer function, $T_3(s)$, contains the required resistors for the VCO tuning range extension. Their ratios are therefore fixed and the filter is realized subject to this constraint. An appropriate capacitor across the VCO, as specified in requirement g, is also available.

The transfer function, $T_3(s)$, assumes the form

$$T_3(s) = \frac{\dfrac{\sigma_1 \sigma_2}{k}}{(s + \sigma_1)(s + \sigma_2)} \tag{70}$$

where σ_1 and σ_2 are real constants and k is the ratio already specified for the VCO tuning-range extension.

V_2/V_1 will be equal to $k_{DC} F(s)$. Also it is found quite readily that $T_1(s) = 1/R_a(-y_{21n})$. Hence

$$\frac{V_2}{V_1} = k_{DC} F(s) \frac{T_2(s)\, T_3(s)}{R_a(-y_{21n})}$$

$$= \frac{k_{DC} a_o b_o}{(s^3 + a_2 s^2 + a_1 s + a_o)(s^2 + b_1 s + b_o)} \tag{71}$$

the denominator of F(s) has been expressed as the product of a cubic and quadratic.

If

$$T_2(s) = \frac{b_o}{s^2 + b_1 s + b_o} \quad ,$$

(72)

then it follows that

$$R_a(-y_{21n}) = \frac{\sigma_1 \sigma_2}{k_{DC} k a_o} \frac{s^3 + a_2 s^2 + a_1 s + a_o}{(S + \sigma_1)(s + \sigma_2)} \quad .$$

(73)

Now since $F(0) = 1$,

$$k_{DC} = \frac{T_2(s) T_3(s)}{R_a(-y_{21n})}\bigg|_{s=0} = \frac{R_o}{k R_a} \quad ,$$

(74)

substituting in Eq. (73)

$$R_o(-y_{21n}) = \frac{\sigma_1 \sigma_2}{a_o} \frac{s^3 + a_2 s^2 + a_1 s + a_o}{(s + \sigma_1)(s + \sigma_2)} \quad .$$

(75)

Equations (70), (72) and (75) define transfer functions realizable by the structure in Fig. 23.

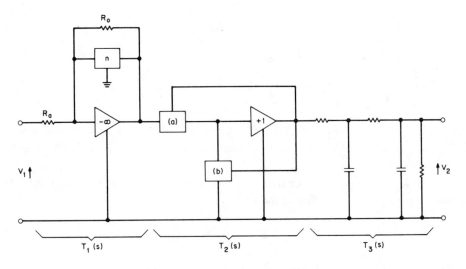

Fig. 23. Possible structure for loop filter.

The actual realization is reasonably straightforward and is not discussed here; however, the element values can be derived as explicit functions of the coefficients [Eq. (76)]. The detailed structure is shown in Fig. 24.

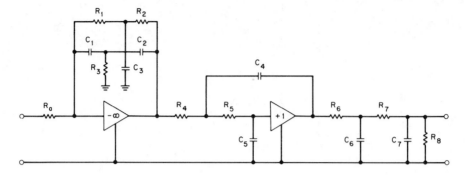

Fig. 24. Detailed structure of loop filter.

$$R_1 = R_2 = 0.5 \qquad\qquad C_1 = C_2 = 2a_2/a_1$$

$$R_3 = a_1^2/4a_0 a_2 \qquad\qquad C_3 = 4/a_2$$

$$R_4 = R_5 = 0.5 \qquad\qquad C_4 = 4/b_1$$

$$R_6 = g(k-1)/(g+1) \qquad C_5 = b_1/b_0$$

$$R_7 = (k-1)/(g+1) \qquad\quad C_6 = 2(g+1)^2/g(k-1)\left(a_2 + \frac{a_0}{a_1}\right)$$

$$R_8 = 1.0 \qquad\qquad\qquad\quad C_7 = s(k+g)/(k-1)\left(a_2 + \frac{a_0}{a_1}\right) \tag{76}$$

where

$$g = \left(\frac{a_1 a_2}{a_0} + \frac{a_0}{a_1 a_2} - 2\right)\frac{k}{4} \quad . \tag{77}$$

Equation (77) always gives a positive value of g except where $a_1 a_2 = a_0$ at which point g = 0. This, however, is a very unlikely event. If it does occur a different realization has to be found.

The value k is always larger than 1, since it represents the attenuation ratio of the pretune extension. In terms of the elements of Fig. 23:

$$k = 1 + \frac{R_6 + R_7}{R_8} \quad . \tag{78}$$

Formulae (76) are easy to use once the function is given.

If for some reason, active realizations are unacceptable, the filter can readily be realized in passive form. Once the required transfer function is known, the filter realization in resistively terminated L, C form follows conventional passive network synthesis procedures.

G. CONCLUSIONS

The Digiphase principle, although capable of arbitrary resolution, cannot, in practice, provide high spectral purity and wide bands at high frequencies. The reason is inherent in the cancellation principle used. A canceling ramp has to be very accurately aligned in amplitude and time to get good suppression of unwanted modulation. If the equipment has to work over a number of years without adjustment, it is unrealistic to expect a variation of less than several percent. One percent implies about 40 dB of sideband suppression, so in reality, about 30 dB over a period of years is very optimistic. If higher spectral purity is required, the Digiphase principle, in the form here described, cannot be used. The best compromise is a three-part combination of a basic loop with resolution, f_R, a 2- or 3-bit modulation down to $f_M (f_M = 1/4\ f_R$ or $1/8\ f_R)$ and a fine-resolution digital synthesizer. A schematic of such a system with 2-bit modulation down to $1/4\ f_R$ is shown in Fig. 25. It is assumed that $1/4\ f_R$ is sufficiently above the loop-cutoff frequency that the modulation

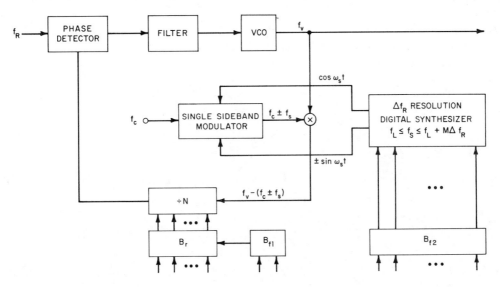

Fig. 25. An optimal PLL synthesizer.

sidebands at $f_v \pm 1/2\ f_R$ and $f_v \pm 1/4\ f_R$ are substantially attenuated. This is usually readily achievable. For example, with a 400-kHz reference frequency, two bits of modulation will result in worst-case sidebands 100 kHz away from the VCO frequency. Also resolution down to 100 kHz has been achieved. A loop filter of 30 or 40 kHz, especially of reasonably high order, can give well in excess of 100 dB of attenuation at 100 kHz while still given quite good settling time (possibly 100-200 μsec). The digital synthesizer will then have to supply 50 kHz of fine resolution if its double side-band capability is used. This requirement is, of course, readily achievable. It will be found that it is not very good practice to have the digital synthesizer go down to DC.

Instead it is preferable to have the lowest frequency produced be above the loop-filter cutoff. Such an arrangement avoids close-in spurious modulation of the VCO output through ground loops, etc. Fortunately, this requirement does not affect the frequency capabilities of the synthesizer.

In the foregoing example, the digital synthesizer could be arranged to supply frequencies in the band 100 kHz to 150 kHz with the required resolution, Δf_R. It is then only a matter of appropriately translating the external frequency command to have the digital synthesizer supply the required sidebands to the single-sideband modulator and to have B_r and B_{f1} compensate for the fact that the fine resolution rides on 100 kHz.

The synthesizer just described can achieve very good fine resolution without any problems with close-in sidebands, which can be suppressed to any desired value, depending on the loop filter. It is also important to note that this can be achieved without relying on very fine balancing of two canceling modulation waves.

Finally, it should be added that, at the time of this writing, progress in the digital IC field has already made a number of implementations described here obsolete. For example, the high speed presettable counter and prescale-by-2 could, in some applications, be replaced by bi-quinary counter. The recent expansion of VLSI in the form of microprocessors will undoubtedly have a significant effect on all digital designs in the near future. Conceivably, most of the digital controls described in this book could be handled by a microprocessor.

REFERENCES

1. Paul Gili, "Getting the Drift of VCO Instability," Microwaves 42-45 (March 1974).

2. W. B. Goggins, "Post-tuning Drift and Noise Properties of VCOs," Microwave J. 34-35 (January 1974).

3. R. Fletcher and M. J. D. Powell, "A Rapidly Convergent Descent Method for Minimization," Computer J. 6, 2, 163-168 (1963).

4. C. J. Byrne, "Properties and Design of the Phase-Controlled Oscillator with a Sawtooth Comparator," Bell Syst. Tech. J. 41, 559-602 (March 1962).

5. A. J. Goldstein, "Analysis of the Phase-Controlled Loop with a Sawtooth Comparator," Bell Syst. Tech. J. 41, 603-633 (March 1962).

6. C. R. Cahn, "Piecewise Linear Analysis of Phase-Lock Loops," Trans. IRE SET-8, 8-13 (March 1962).

7. R. C. Tausworthe, "New Calculation of Phase-Locked Loop Performance," JPL Sect. Rep. 8-583 (December 16, 1959).

8. C. S. Weaver, "A New Approach to the Linear Design and Analysis of Phase-Locked Loops," Trans. IRE SET-5, 166-178 (December 1959).

9. A. F. Evers, "A Versatile Digital Frequency Synthesizer for Use in Mobile Radio Communication Sets," Electronic Engineering 28, 296 (May 1966).

10. P. J. Rasch and J. F. Duval, "A High Speed Microwave Frequency Synthesizer," Microwave J. 9, 97-101 (June 1966).

11. R. K. Keeman, "An Introduction to and Survey of Frequency Synthesizer Techniques," Natl. Telemetering Conf., Houston, Texas (April 1968).

12. B. C. Welling and E. L. Renschler, "Design of a Phase-Locked Loop Multi-Channel Digital Frequency Synthesizer," Proc. NAECON (1968).

13. V. T. Pivovar, "Digital Frequency Synthesizer," Pribory i Sistemy Automatiki 9 (1969).

14. E. L. Renschler and B. C. Welling, "Frequency Synthesizing with the Phase-Locked Loop," Electronic Engineer 29, 84-90 (June 1970).

15. L. F. Blachowicz, "Dial Any Channel to 500 MHz," Electronics 39, 69-80 (2 May 1966).

16. F. M. Gardner, Phaselock Techniques (Wiley, New York, 1966).

17. A. J. Viterbi, Principles of Coherent Communication (McGraw-Hill, New York, 1966).

Chapter V
Digital Frequency Synthesizers
J. Tierney

An attractive alternative approach to frequency synthesis is the digital or sampled data technique. Other methods described in this book produce analog or continuous signals directly.[1-5] The digital frequency synthesis approach uses the stable source frequency to define sampling times at which digital sinusoidal sample values are produced. These samples are D/A converted and smoothed in time by some realizable linear filter to produce analog frequency signals (Fig. 1): single frequency (empty dots), lower frequency (solid dots). The samples for both frequencies are produced

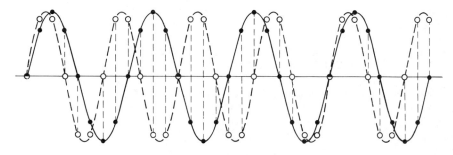

Fig. 1. Digital frequency determination.

at the same sample times determined only by the basic frequency standard and invariant to which frequency is produced. The calculation determines the frequency by varying the amplitude of the samples, not by changing the sampling interval. Digital synthesis consists then of computing at some real-time interval, T, values of a desired phase angle, $\omega_o t = \omega_o n T$, with ω_o a desired synthesizer output frequency, and then using this value of phase angle to compute a sinusoidal output sample, $\sin \omega_o n T$, in real time. Since the phase angle is a linear function of time and treated modulo 2π, a simple accumulator of phase increments, $\omega_o T$, with overflow at effective 2π, solves the angle computation.

Assuming the smoothing filter presents no serious design problem (a reasonable assumption), and the D/A converter produces distortion-free analog samples (not so reasonable), then a suitable technique for digital sample value determination remains as the design problem.

Methods available to determine the sine and cosine of some argument, $n\omega T$, where n is a sample index, T the sampling interval, and ω the desired frequency, are basically:

1. Digital recursion oscillator
2. Direct computation based on some numerical approximation
3. Direct table look-up.

$\omega_o t = \omega_o n T$

$\sin \omega_o n T$

A. DIGITAL SAMPLE DETERMINATION

The digital recursion oscillator[6] is an obvious first choice for sinusoidal sample output, which means a difference equation realization whose Z transform has poles on the unit circle. By starting such a recursion with the proper initial conditions, sinusoidal samples are produced (Fig. 2).

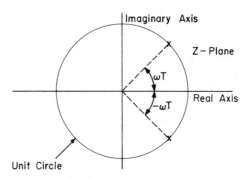

Fig. 2. **Pole position of digital recursion for sinusoidal samples:** $y_n = (2 \cos \omega T) y_{n-1} - y_{n-2}$; $y_0 = \cos \phi$; $y_{-1} = \cos \phi - \omega T$.

There are at least two problems with this approach. The noise produced (i.e., the output sample is not a perfect sinusoidal sample) can grow in size with the number of recursion cycles until a limit cycle[7] occurs or some kind of saturation clamping causes a nonlinear steady output. In most cases, the output signal-to-noise ratio (SNR) from such an oscillator is unacceptable unless the recursion is reset after a certain number of iterations. Under certain conditions of coefficient setting and round-off iteration, the recursion may produce a limit cycle oscillation that is pure enough. However, the behavior of these recursive limit-cycle oscillators is not yet well enough understood to produce reliable oscillators over a wide range of frequencies.

Another problem with the recursion approach involves $2 \cos \omega T$, the frequency determining coefficient. Note that a frequency of ω requires a cosine function coefficient setting. The sinusoidal function is needed to set an oscillator to provide the sinusoidal function. The oscillator is not the complete solution after all. In addition, the problem of representing the $2 \cos \omega T$ coefficient with a finite-length word in a digital computation means that the frequency setting may produce a different frequency from that desired; in effect, a limitation on time-frequency steps available from the synthesizer.

The other two methods for sine and cosine generation are similar. By <u>numerical approximation</u>, a small number of stored constants characterize some approximation to the desired function. Each time the function is determined (sinusoidal sample output) a computation based on the stored constants is carried through. The length of the computation depends on the error allowed (really the SNR allowed) and the efficiency of the approximation.

By direct table look-up, the computation is minimized at the expense of stored constants. An effective approach is really a combination of table look-up and computation that can be thought of as either one if minimum and maximum stored coefficents are considered. The decreasing cost of read-only memory (ROM), compared to other digital components, is tending to drive future implementations toward direct table look-up.

B. MODIFIED TABLE LOOK-UP

Consider computing values of:

$$\cos 2\pi fnT = \text{Re } \exp[j2\pi fnT]$$

$$\sin 2\pi fnT = \text{Im } \exp[j2\pi fnT]$$

where $f = kf_o$, f_o = lowest frequency, n = time index, and T = sample interval. That is, multiples of some lowest frequency can be computed

$$f_o = 1/NT$$

where N is a design parameter. Then the exponential becomes

$$\exp[j2\pi fnT] = \exp[j\frac{2\pi}{N} nk] \quad .$$

Samples of this complex exponential are equivalent to values of sine and cosine of the argument $2\pi fnT$ with

$$f = \frac{k}{NT} \quad .$$

Computing samples of this complex exponential indexed on a frequency index k, and a time index n, is equivalent to computing coordinates of one of N equispaced points around the unit circle (Fig. 3) in the complex plane described by.

$$\exp[jo] \, , \, \exp[j\frac{2\pi}{N}] \, , \, \exp[j\frac{2\pi}{N} 2] \, \dots \, . \, \exp[j\frac{2\pi(N-1)}{N}] \quad .$$

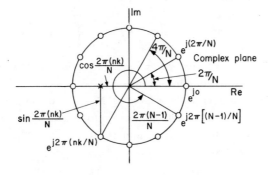

Fig. 3. Equispaced samples on the unit circle.

For a particular frequency index k, the argument of the exponential varies in increments of $(2\pi/N)$ k in successive time indices. The product nk is treated modulo N since $\exp[j(2\pi/N)(X + N)] = \exp[j(2\pi/N)X]$ for any X. The generation of samples of a complex sinusoid consists then of accumulating multiples of k [i.e., nk at time n, $(n + 1)$k at time n + 1], and using the accumulated value to calculate $\exp[j(2\pi/N)$ nk]. If in accumulating multiples of k, the accumulator is not initially zero, but contains some constant C, the argument of the exponential becomes $[j(2\pi/N)(nk + C)]$ which affects the phase of the result, but not the frequency. This can be useful in phase control.

To determine the value $\exp[j(2\pi/N)(nk + C)] = \exp[j(2\pi/N)Y]$ consider the simplest case, namely, a table storing N values from $\exp[j(2\pi0/N)]$ to $\exp j[2\pi(N - 1)/N]$. The computation then consists of using the value in the accumulator, Y, to index the table of N values producing a single complex sample.[8] As Y increases the table is scanned, producing sinusoidal samples. For a larger k value the table is covered faster (the interval between successive Y's, or nk's is larger). Such an approach is suitable for small values of N, but usually large N (a large set of frequencies) is of interest so that a single table look-up becomes impractical because of the size of the table.

For large N and $0 \leqslant Y \leqslant N - 1$, Y may be broken into a sum of several words, each of which represents a part of Y. If Y = q + r + s, then

$$\exp[j\,\frac{2\pi Y}{N}] = \exp[j\,\frac{2\pi(q + r + s)}{N}] = \exp[j\frac{2\pi q}{N}]\,\exp[j\frac{2\pi r}{N}]\,\exp[j\frac{2\pi s}{N}]$$

where each factor takes on many fewer than N values and the overall storage has been reduced. For example, for N a power of 2, say 2^b, then $0 \leqslant Y \leqslant 2^b - 1$, and can be represented as a binary number, b digits long. $Y = \alpha_0 2^0 + \alpha_1 2^1 + \alpha_2 2^2 + \ldots + \alpha_{b-1} 2^{b-1}$. Factor the exponential into b factors, each of which is of the form $\exp[j(2\pi\alpha_i 2^i/N)]$ with α_i = 0 or 1. Thus the table is reduced from 2^b complex entries to b complex entries $(\log_2 2^b)$, and $(b - 1)$ complex multiplications are needed to obtain the value, $\exp[j(2\pi Y/N)] = \exp[j(2\pi/N)(nk + C)]$. Obviously, the number of factors or the number of complex multiplies is one of the design parameters in this approach to frequency synthesis.

For purposes of this approach, factoring the complex exponential into two terms allows for a very efficient use of ROM. In addition, advantage may be taken of approximations to the sine and cosine of small angles as well as symmetries in the functions to effect further savings in read-only storage.

Some generalizations about storage and multiplication optima are appropriate before moving to the design of a complete frequency synthesizer.

If the number of stored constants is fixed at N, then for no multiplies, there are 4N values around the unit circle. This assumes only a quarter-sine function table need be stored. These 4N values around the unit circle yield N distinct frequencies if the highest frequency is limited to one fourth the sampling frequency. If the

N stored values are split into two sections and one complex multiply is used to produce output samples, the maximum number of available frequencies is

$$F = (N/2)^2 \quad .$$

It is easy to show that splitting N into equal parts optimizes the number of frequencies. In general, for n complex multiplies

$$F = (\frac{N}{n+1})^{n+1} \quad .$$

Maximizing this expression for the number of distinct frequencies with respect to n, the result is

$$n + 1 = N/e \quad \text{or} \quad n = (N/e) - 1$$

and $F_{max} = e^{N/e}$ with e the natural logarithm base. For example, if N = 32, $F_{max} = e^{32/e} \approx 2^{17}$ and n = 10.8. To implement such an optimum in terms of integer arithmetic, it is necessary to make integer approximations to the continuous arithmetic implied. Instead of splitting the 32 stored values into 11.8 pieces (n + 1), which are combined with 10.8 multiplies (n), consider the 32 values split into ten groups of three each and a single group of two values. The total implementation then requires 10 multiplies and $F_{total} = 3^{10} \times 2 \approx 118,000$, compared to $F_{max} \approx 130,000$ for N = 32 using the unrealizable value, a difference of about 10 percent. The accumulator storing values of nk for this implementation uses binary arithmetic in the least significant digit, ternary arithmetic in its next ten digits, and finally, two end stages of binary arithmetic for quadrant control. Note that each group (Fig. 4) stores unity, which is not really necessary. Complementing for quadrant control would have to consider the combination of binary and ternary arithmetic. If a straight binary implementation were used, the total number of frequencies would be 2^{16}, only a factor of two below the optimum, a reasonable price to pay for ease of implementation.

If the desired number of frequencies is fixed and storage is minimized with respect to the number of multiplies:

1. For a fixed number of frequencies F, and no multiplies, F storage values are needed.

2. For one multiply, $\sqrt{F}$ storage values in each of two groups, so total storage of $2\sqrt{F}$ values are needed.

3. For the general case of n multiplies and n + 1 storage groups:

$$\text{Storage} = (n + 1)(F)^{1/n+1} \quad .$$

$$S_{min} = (F)^{1/\ln(F)} \ln(F)$$

and occurs at $n = \ln(F) - 1$. For a desired $F = e^a$, $S_{min} = ae$. For example, if $F = e^{12} \approx 2^{17}$, $S_{min} = 32.6$ consistent with the previous result with N fixed. If F is chosen as some desired power of 2, say 2^b, then

$$S_{min} = (2^b)^{1/b \ln 2} b(\ln 2) = 2^{1/\ln 2} b(\ln 2)$$

$$= e\, b(\ln 2) = 1.88b \quad .$$

If results of the previous section are recalled for binary implementation of $F = 2^b$, $\log_2 F = b$ stored constants are needed along with $(b - 1)$ multiplies. Actually, 2b stored constants are needed (b of them are simply $e^{j0} = 1$) to be consistent with the results in this section. So the comparison is 1.88b stored constants in the unrealizable optimum versus 2b constants in the binary arithmetic version — a 6 percent increase. For a

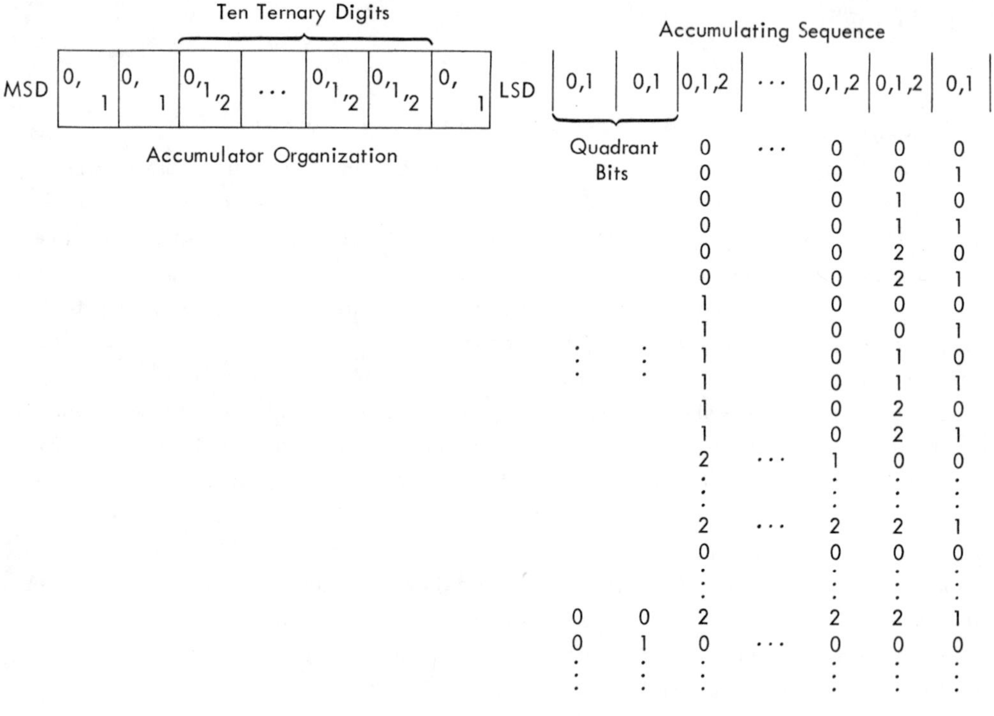

Fig. 4. Storage organization.

Highest ternary digit $\quad e^{j(2\pi/2 \cdot 3)}, \quad e^{j(\pi/2 \cdot 3)}, \quad e^{j0}$

Next highest $\quad\qquad e^{j(2\pi/2 \cdot 3^2)}, \quad e^{j(\pi/2 \cdot 3^2)}, \quad e^{j0}$

$\vdots$

Lowest ternary digit $\quad e^{j(2\pi/2 \cdot 3^{10})}, \quad e^{j(\pi/2 \cdot 3^{10})}, \quad e^{j0}$

Least significant bit $\quad e^{j(\pi/2 \cdot 3^{10} \cdot 2)}, \quad e^{j0}$

base 3 example of $F = 3^b$, and a base 3 implementation, the optimum is only about 0.6 percent better in terms of stored constants required. In spite of this "optimality" of base 3 implementation, binary realizations are certainly preferred for obvious practical reasons.

C. COMPLETE DIGITAL SYNTHESIZER

In a digital synthesizer that produces quadrature outputs (Fig. 5) an input frequency control word is stored in a register and used to update an accumulator every T sec. This is the phase argument of the sinusoidal computation. Every T-sec interval, determined by the clock or stable frequency source in the system, k is added

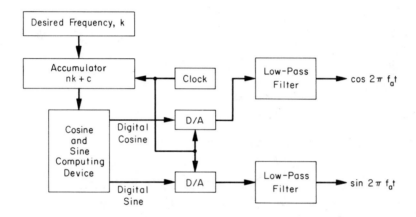

Fig. 5. Digital synthesizer that produces quadrature outputs.

to the present contents of the accumulator register to produce the new value, nk + C = Y. Each time a new value is determined it is used to compute the real and imaginary parts of $e^{j(2\pi/N) Y}$ by one of the methods proposed. The length of the accumulator register determines the number of distinguishable points around the unit circle, and therefore, the size of the frequency set the device is capable of producing. The cosine and sine computing device produces the digital sample value as determined by the accumulator register. If the computation is implemented as a modified table look-up, then the total effective table must correspond to the accumulator size. Note that the accumulation and overflow process for the indexing nk, or the computing device arithmetic, is not restricted to any particular number system. The operations are completely general, as indicated in the previous section. In fact, for most applications, a choice of binary arithmetic is the most appropriate one. Under certain conditions, when the synthesizer is under control from an external source (i.e., the frequency input word k) the number representation for k may be in other than fixed-point binary, so that the conversion time from the original k number representation (outside world) to the synthesizer accumulator arithmetic may be unacceptable. If so, the accumulation, and even the computing, may be done in the outside system's

arithmetic even at a cost of total frequency number or storage size. This design freedom is available.

The computing device outputs (Fig. 5) — $\sin(2\pi/N)$ Y, $\cos(2\pi/N)$ Y — drive a pair
of D/A converters of the proper word length to produce analog samples that are inter-
polated by the output smoothing filters. Before smoothing, the output spectrum of
a T-second sampled sine or cosine would look like that shown in Fig. 6(a). The Nyquist
condition permits production of frequencies just less than $1/2T$ that can be recovered

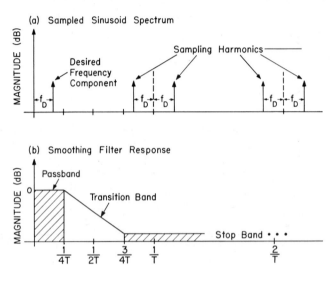

Fig. 6. Spectrum of (a) sampled sinusoid and (b) smoothing
filter.

with ideal lowpass filters with $1/2T$ cutoffs. However, for filtering ease, consider
using only one fourth the sample frequency as the band limit. Such an output smoothing filter (a) passes all frequencies up to $1/4T$ with some design ripple, (b) has a
transition band in the $1/4T$ to $3/4T$ interval, and (c) has some out-of-band attenuation
depending on the sample harmonics allowed [Fig. 6(b)]. For an accumulator that overflows at some N and sample interval T, a digital frequency synthesizer can produce
a low frequency of $1/NT$, and a high frequency of $1/4T$, for a total of $N/4$ different
frequencies (Fig. 4). By taking advantage of the quadrature outputs from this realization the $1/4T$ bandwidth can be doubled by modulating a carrier to $\pm(1/4T)$ for a total
of $N/2$ frequencies. Note that a single output synthesizer can be implemented using
only one D/A converter, if so desired. There is no constraint to produce quadrature
outputs, although they may be desirable in some cases.

D. TWO DESIGN EXAMPLES

Consider two designs using binary arithmetic and standard binary logic. Specifications for the first design are:

2^{15} frequencies

409.6-kHz bandwidth

12.5-Hz frequency spacing (or lowest frequency)

70-dB spectral purity (ratio of power in desired frequency to power in any other 100-Hz band).

If the quadrature outputs are assumed to obtain a bandwidth of $1/2T$, then $T = (1/2 \times 409.6 \text{ kHz}) \cong 1.22$ μsec. In addition, $N/2 = 2^{15}$, or $N = 2^{16}$ so that the accumulator is a 16-bit binary register, and the frequency spacing of $1/NT = 12.5$ Hz. If quadrature outputs are desired, the synthesizer must produce values of $\exp[j(2\pi/2^{16})\,Y]$, $0 \leqslant Y \leqslant 2^{16} - 1$, i.e., one of 2^{16} points equispaced around the unit circle. Since the highest frequency allowed is $1/4T$ (four points around the circle) the largest k value corresponding to this value of frequency is $2^{16}/4$ or 2^{14}. A 2's complement negative frequency input causes samples to occur in the opposite sense around the unit circle, which is meaningful as a negative frequency only if quadrature outputs are available.

The remaining problem is computation of the samples. The computation must be carried to about 12 bits to meet the design requirement of 70-dB purity. That is, if a computation sample consists of a sign and 11 binary digits, and the accuracy is ± the last digit or $\pm 2^{-11}$, a worst-case harmonic caused by such an error will be 66 dB down from the output. (This bound is quite pessimistic; 70 dB is more likely.) Using this 12-bit arithmetic and a single complex multiply with two tables, the following procedure is used.

First, consider the effects of neglecting the least significant bit in the 16-bit index accumulator. Such a move causes an amplitude error of no more than $\sin(2\pi/2^{16})$ which is less than 2^{-12}. The error generated when neglecting this bit (Fig. 7) is always bounded by this amount, even if an infinite number of samples are produced between two values in the table by repeating values. This is equivalent to extending the accumulator register and the frequency control word on the low significance end and ignoring these extra bits in the computation. This implies generating finer and finer frequency steps with a bounded amplitude error (noise) by adding only to the accumulator and not to the memory. This fine frequency generation is analogous to the variable modulus divider technique used in phase-locked loop synthesizers.

The repeating-sample concept can be used to make sure more memory is not used than necessary. Having decided on the accuracy of the output digital sample needed to meet the noise specifications (11 bits and a sign bit), it is necessary only to divide the unit circle into fine enough increments so that $\sin 2\pi/N$ is smaller than the error allowed in the design (12 bits, or 2^{-12}). If the required frequency set is larger than this N implies, then the accumulator is extended, as explained, until the fine increment and the frequency set is adequate. This guarantees maximum memory use. If, on the other hand, the noise requirements are very tight, and the required frequency set sparse, only enough points are used around the circle to give the required frequency set without repeating. In this case, repeating will not produce sufficiently pure samples.

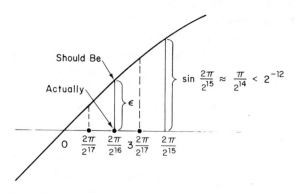

Fig. 7. Ignoring bit 16 means $\sin 2\pi/2^{16}$ is approximated by $\sin 0$. Interpolating many samples between 0 and $2\pi/2^{15}$ as $\sin 0$ means error is less than $\sin 2\pi/2^{15} < 2^{-12}$.

For this design, the least significant accumulator bit is neglected and the sine and cosine of multiples of $2\pi/2^{15}$ must be determined. To divide the required table into two sections, consider factoring the contents of the accumulator with Y represented in binary form

$$Y = 2^0 d_0 + 2^1 d_1 + \ldots + 2^6 d_6 + 2^7 (d_7 + 2d_8 + \ldots 2^7 d_{14})$$

$$= 2^0 d_0 + 2^1 d_1 + \ldots + 2^5 d_5 - 2^6 d_6 + 2^7 (d_6 + d_7 + 2d_8 + \ldots + 2^7 d_{14})$$

$$= f + 2^7 e$$

where

$$f = (2^0 d_0 + 2^1 d_1 + \ldots + 2^5 d_5 - 2^6 d_6)$$

$$e = d_6 + d_7 + 2d_8 + \ldots 2^7 d_{14} \quad .$$

The seventh bit (d_6) has been effectively added and subtracted from the accumulator value, so that

$$e^{j(2\pi/N)Y} = e^{j(2\pi/2^{15})(f+2^7 e)} = [e^{j(2\pi/2^{15})f}] [e^{j(2\pi/2^8)e}] \quad .$$

The computation of the complex exponential is then reduced to two table look-ups,

$$e^{j\frac{2\pi}{2^{15}}f} \quad , \quad e^{j\frac{2\pi}{2^8}e} \quad .$$

and a complex multiply. Index e consists of the eight high-order bits of the accumulator rounded by the bit d_6; while index f consists of the six lower bits, d_5 through d_0, if d_6 is zero, or these bits are 2's complemented if d_6 is one. From the view of computing the value of a point at a particular angle, $(2\pi/N)$ Y, around the unit circle (Fig. 8), the value of e determines which of 2^8 equally coarsely spaced points is nearest the desired point, and the value of f determines which of 64

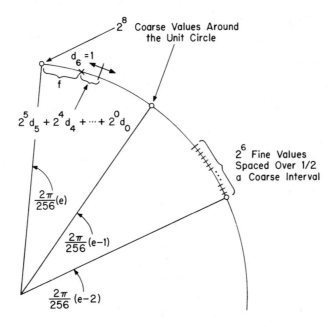

Fig. 8. Computing the value of a point at a particular angle.

possible angular corrections should be added to or subtracted from the coarse points to get the desired value. The angular correction is, of course, a complex multiplication. For this two-factor approach, the complex multiply implements the trigonometric identities

$$\sin(x + y) = \sin x \cos y + \cos x \sin y$$

$$\cos(x + y) = \cos x \cos y - \sin x \sin y$$

with $x = (2\pi/2^8)$ e, $y = (2\pi/2^{15})$ f where f may be positive or negative (according to bit d_6).

To compute the value at X (Fig. 8) the eight high-order bits are augmented by one (since $d_6 = 1$) yielding e, and the value of f is the 2's complement of the distance above the center of the coarse interval. In other words, go to a larger angle and decrement because $d_6 = 1$. If $d_6 = 0$, start at the lower angle and increment.

The bit d_6 breaks the coarse intervals in half. If this bit is one, the coarse value must be slightly larger than the desired value. The fine correction is then clockwise. If d_6 is a zero, the coarse value just below the desired value will do, and the correction is counterclockwise.

The cosine component of $\exp[j2\pi f/32768]$ is between 1.0 and 0.9999247, a difference in the 14th bit of its binary representation. Therefore, approximate it by 1. The sine component is so small that its six most significant bits, not counting the sign bit, are equal to the sign bit, which, in turn, is equal to d_6. Thus the ROM indexed by f need only save the five least significant bits of the 11-bit plus sign representation of the sine corresponding to each of the 64 positive values of f; the sines corresponding to negative values of f can be found by taking the 2's complement of f and changing the sign of the result. It is easier to take the 1's complement, and this also results in an insignificant error.

The value of $\exp[j2\pi e/256]$ can also be found in a table with only 64 values corresponding to a quarter cycle of a sine wave. The sine and cosine components are addressed with the six least significant bits of e, and its 2's complement; and the two most significant bits of e are used to exchange the components and complement either or both, if necessary; mathematically, if the six least significant bits of e are g and the two most significant bits are h, look up $\sin 2\pi(64 - g)/256 + j \sin 2\pi g/256$ and multiply the result by j^h.

The final operation is multiplication of the coarse estimate $\exp[j2\pi e/256]$ by the fine corrector $(1 + j \sin 2\pi f/32768)$. This requires two 8-by-5 bit multiplications and two additions. Two answers are kept to an accuracy of 11-bits plus sign and fed to the D/A converters.

The ROM requirements are 64 words of 5-bits, fine angle, and 11-bits, coarse angle, that can be combined into a 64-word by 16-bit memory that is accessed three times in a computation (Fig. 9). The basic sampling interval or computation interval is divided into three separate memory indexing times. During the first index time the word f or its 2's complement is used to look up the fine corrector, $\sin(2\pi f/2^{15})$, which is stored in a 5-bit register in the array multiplier for the entire time T. In addition, the two 12-bit registers feeding the D/A converters are cleared. During the second index time the coarse $\sin(2\pi e/2^8)$ is indexed from the coarse word e (actually the lower 6 bits of e, or their 2's complement) and stored in the right-hand, 12-bit register that was cleared previously. In addition, the product out of the array multiplier is $\sin(2\pi f/2^{15}) \sin(2\pi e/2^8)$ which is stored in the left-hand, 12-bit register cleared previously. At this point, $\sin \times \sin$ is in the left-hand register and sin (really sin times cos) is in the right-hand register. The coarse $\cos(2\pi e/2^8)$ is indexed, which again goes to the array multiplier and directly to the right-hand register. The right-hand register stores the $\cos(2\pi e/2^8) \pm$ the contents of the left-hand register, $\sin(2\pi f/2^{15}) \sin(2\pi e/2^8)$, because of the connections between opposite registers and adders. In a similar fashion, the left-hand register stores $\sin(2\pi f/2^{15}) \cos(2\pi e/2^8) \pm \sin(2\pi e/2^8)$, so that left and right registers end up with cosine and sine outputs, respectively. The direct connection of registers to opposite adders saves some unnecessary gating and wiring.

The 5×8 array multiplier need not be any larger because the 5-bit $\sin(2\pi f/2^{15})$ has only five low-order significant bits out of 11 so that keeping an 11-bit and sign word requires only this amount of input significance. 4-bit standard TTL adder

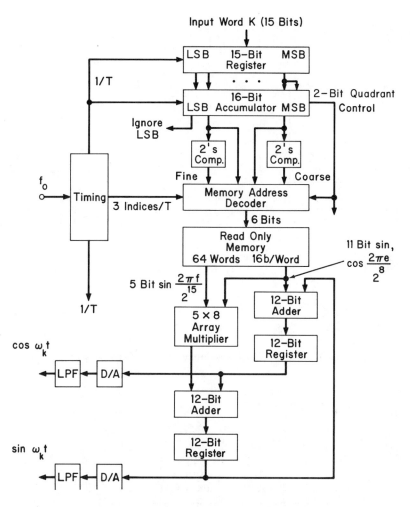

Fig. 9. First of two digital frequency synthesizer designs that use binary arithmetic and standard binary logic.

packages implement the array (Fig. 10). Note that partial products have been inter-connected to minimize overall delay time. The 5 × 8 multiply requires about 160 nsec.

The two 12-bit registers drive D/A converters (Fig. 9), which, in turn, drive two lowpass smoothing filters. These filters, designed as 5th-order elliptic function filters with 0.18-dB in-band ripple and 80-dB out-of-band rejection, provide the filter realization and frequency response shown in Fig. 11.

A synthesizer designed and built in 1970 according to the foregoing discussion, requiring about 85 standard TTL packages including ROM, dissipates about 12 W.

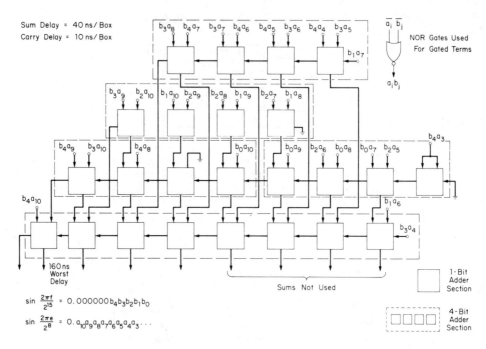

Sum Delay = 40 ns / Box
Carry Delay = 10 ns / Box

NOR Gates Used
For Gated Terms

160 ns
Worst
Delay

Sums Not Used

1-Bit
Adder
Section

4-Bit
Adder
Section

$$\sin \frac{2\pi f}{2^{15}} = 0.000000\, b_4 b_3 b_2 b_1 b_0$$

$$\sin \frac{2\pi e}{2^8} = 0.\, a_{10} a_9 a_8 a_7 a_6 a_5 a_4 a_3 \cdots$$

Fig. 10. 4-bit TTL adder packages implement the 5 × 8 array multiplier.

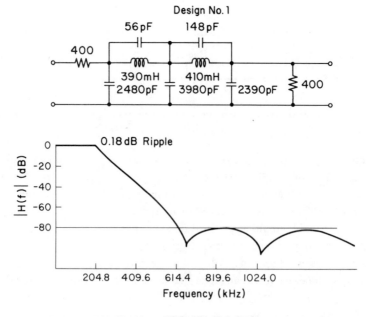

Design No. 1

Fig. 11. Smoothing filter realization.

Specifications for the second design are:

High frequency	1.0 MHz
Lowest frequency, fo	0.01 Hz (and frequency resolution)
SNR	60 dB (ratio of desired output frequency to noise in 100-Hz band)

Single sine out (no quadrature).

Design decisions must again be made concerning:

Sampling interval T

Total number of distinct computed points in the sine wave period and length of indexing accumulator register

Length of binary words in the arithmetic for the required accuracy or SNR

To determine interval T, consider the sampled spectrum discussed previously (Fig. 6). For a high frequency of 1 MHz as a design requirement and for ease of filtering, the highest generated frequency is $1/4T$, the sampling interval is $1/4T = 10^6$ or T = 250 nsec, which is an excessive trade. 250 nsec does not allow any margin for a TTL-implemented computation of sine. It would make more design sense to make the filter cutoff rate somewhat sharper and ease off on the sampling (or computation) interval. Working backward, a more reasonable sampling interval of 400 nsec (2.5-MHz sampling frequency) can be chosen. Then the highest allowable frequency (1 MHz) implies a transition band for a smoothing filter from 1 to 1.5 MHz and a stop band starting at 1.5 MHz — a cutoff ratio of 1.5 to 1.0. A tighter filter than the 3 to 1 design for the previous smoothing is still a straightforward realization, so that T = 400 nsec.

From the lowest frequency requirement of 0.01 Hz, $1/NT = 0.01$ Hz can be set up to determine the length of the indexing accumulator

$$\frac{1}{N400 \times 10^{-9}} = 10^{-2} \quad , \quad N = \frac{1}{400 \times 10^{-11}} = \frac{10^{11}}{400} = 25 \times 10^7 \quad .$$

So N has been determined, but what does this number mean for a binary realization accumulator? It means the accumulator is cleared before it overflows, and more importantly, one-quarter sine wave storage cannot be used for the most significant bits because decoding no longer occurs with the highest two bits. There are several choices:

1. Implement the accumulator in decimal arithmetic and do the table look-up computation in modified decimal (i.e., decimal storage indexing, but binary computation).

2. Adjust the sampling rate slightly so that N becomes a power of 2 and the accumulation is modulo a power of 2 so that the high order 2 bits are quadrant bits and only 1/4 coarse sine table need be stored. The frequency control word for this case can be a binary input from $00\ldots 1$ representing 0.01 Hz to the binary equivalent of 10^8 representing 1 MHz.

3. Using the foregoing technique, the device can be controlled from a BCD word since the frequency increment of 0.01 Hz lends itself to BCD control words. In other words, if the synthesizer is a lab instrument, a BCD control is probably best; otherwise, the direct binary word control is fine. If a BCD frequency word is used, then a BCD-to-binary conversion must be implemented.

Again, one has a choice of methods from a slow serial implementation to a fast array of adders coupled to each BCD digit. The method used depends on the control requirements. For this design, the technique described in method 2 is preferred implying a binary control word input. Implementing this:

$1/NT = 0.01$ Hz for T around 400 nsec and N around 250,000,000. The next greater power of 2 is $2^{28} = 268,427,456$ and the corresponding $T = 372$ nsec. Using $T = 372$ nsec and $N = 2^{28}$ or an accumulator 28 bits long, consider the -60 dB noise requirement in terms of the number of distinct points to be computed. A noise of -60 dB implies 10-bit (9 bits plus a sign) output words from the synthesizer. If computed samples are spaced in angle to keep the error less than 2^{-10}, then

$$\sin \frac{2\pi}{N} < 2^{-10} \quad \text{or} \quad N \approx 2\pi \, 2^{10} \approx 2^{13}$$

and only the most significant 13 bits of the 28-bit register are needed to compute 2^{13} distinct sine samples. The fine points associated with a 28-bit accumulator are obtained by repeating one of the 2^{13} points.

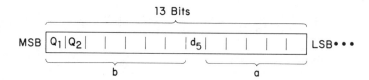

Fig. 12. Accumulator partitioning.

Having decided on T, accumulator length, and the total number of distinct points needed in the computation, all that remains is a suitable technique using the 13 most significant bits of the accumulator to index a pair of ROMs. In other words, how much smaller memory than 2^{11} words of 9 bits each can be arrived at using two tables and one real multiply (want only one sine, not quadrature output)? Consider the accumulator split as shown in Fig. 12.

If the accumulator's 13 bits is considered a binary word

$$Y = d_0 2^0 + d_1 2^1 + \ldots + d_{12} 2^{12} \quad ,$$

a slightly better split can be used than that in design 1:

$$Y = d_0 2^0 + d_1 2^1 + d_2 2^2 + d_3 2^3 + d_4 2^4 + d_5 2^5 + 2^6(d_6 + d_7 2 + \ldots + d_{12} 2^6)$$

$$= d_0 2^0 + d_1 2^1 + \ldots + d_5 2^5 - 2^5 + 2^6(1/2 + d_6 + d_7 2^7 + \ldots d_{12} 2^6)$$

$$= d_0 2^0 + d_1 2^1 + \ldots + (d_5 - 1) 2^5 + 2^6(1/2 + d_6 + d_7 2^7 + \ldots + d_{12} 2^6)$$

$$= d_0 2^0 + d_1 2^1 + \ldots - d_5 2^5 + 2^6(1/2 + d_6 + d_7 2^7 + \ldots + d_{12} d^6)$$

$$= a + 2^6 b$$

where

$$a = d_0 2^0 + \ldots - \bar{d}_5 2^5$$

$$b = 2^6(1/2 + d_6 + \ldots + 2^6 d_{10}) \quad .$$

In splitting the accumulator bits of interest into two words, bit 6, namely (d_5), is used as a dividing bit for the coarse interval defined by the word b. By adding the term 1/2 in the coarse word, the coarse words are shifted around the unit circle by 1/2 a coarse interval and the need to add a low significance bit to the coarse word is eliminated as was done in design 1. Once again a coarse word is looked up and a fine correction added or subtracted depending on bit d_5. Implementing

$$\sin(a + 2^6 b) \frac{2\pi}{2^{13}} = \sin \frac{a2\pi}{2^{13}} \cos \frac{b2\pi}{2^7} + \sin \frac{b2\pi}{2^7}$$

where $\cos(a2\pi/2^{13}) \approx 1$ for 9 bit and sign representation and $\sin(a2\pi/2^{13})$ needs to be represented as four bits since only the first five bits after the point are zeros (i.e., $\sin(2^5/2^{13}) 2\pi = \sin(\pi/2^7) = 0.00000\ 1100)$. The total memory requirement is 32 words 9 bits/word and 32 words 4 bits/word, or 32 words 13 bits/word total = 13×2^5 bits compared to a full table look-up of 9×2^{11}. So the split into two tables cost a modest multiply while saving storage. If more multiply time could be afforded, storage could be reduced even more.

The 28-bit accumulator in the 10-bit-output, single component, frequency synthesizer (Fig. 13) generates the linear index function nk, whose most significant 13 bits is used as angle information. At any nk time, index twice. The first index provides the fine-correction sine and coarse cosine that are stored in the array multiplier. The second index time indexes the coarse memory for sine that is added to the product coming out of the array. So at the end of two index times,

$$\sin \frac{2\pi a}{2^{13}} \cos \frac{2\pi b}{2^7} + \sin \frac{2\pi b}{2^7}$$

is stored in the output register where each term has the correct sign from looking at Q_1, Q_2 and d_5 and using the 2's complement and 1's complement when indicated. This design requires about 50 standard TTL packages including ROM and dissipated about 7 W.

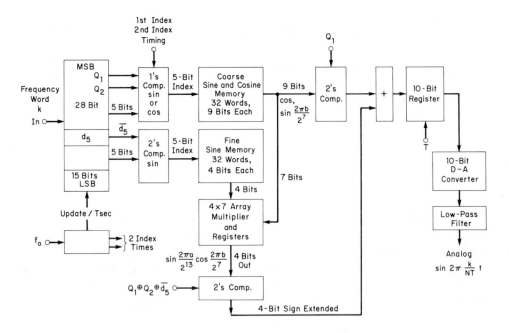

Fig. 13. Second of two digital frequency synthesizer designs.

E. OUTPUT NOISE AND SWITCHING TIME

Noise in the synthesizer output signal is caused by three internal effects plus the external frequency standard used as the timing source. The internal sources are (a) truncation in the sine and cosine computation, (b) distortion caused by D/A effects, and (c) sampling harmonics passing through the output smoothing filter. Phase noise on the external frequency standard is responsible for jitter in the final sample outputs and causes a corresponding output phase noise. Starting with phase noise, each internal noise is considered:

1. Source Phase Effects

The frequency source defines sampling times for the digital synthesizer, signifying interest in zero crossings and use of the source after hard limiting. Obviously, the jitter in these zero crossings represents the phase or frequency modulation in the source. There is a great deal of jitter in the synthesizer's internal logic because of noise pickup and variations from logic element to element. This jitter is of no concern if maximum delays are not exceeded and computation is finished before its sample is needed. As long as the source is used (Fig. 14) a perfect source (zero phase modulation)

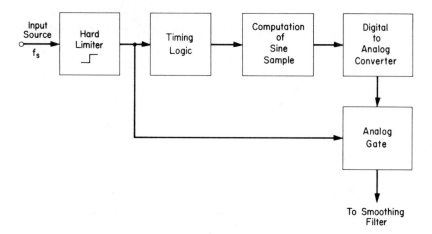

Fig. 14. **Phase noise control circuitry.**

yields zero phase jitter samples, assuming zero gate jitter. Any phase jitter on the hard-limited source shows up as sample jitter on output. If the hard-limited signal does not drive an analog gate, but drives the D/A directly, then the D/A timing jitter introduces extra phase modulation. It is important to see that the source only introduces jitter on output samples reflecting its own zero crossing jitter regardless of what happens internal to the digital logic.

2. Truncation Noise

Since the calculation for a particular sample (i.e., a particular value in the index accumulator) is always the same, the output samples are those of an exact sinusoid with some deterministic noise sample train added to it due to truncation. For an arbitrary generated frequency the truncation noise has a period equal to NT, the largest period possible, so that the truncation noise consists of a line spectrum with frequency spacing $f_0 = 1/NT$. If the generated frequency k/NT has one or more factors of 2 in k, i.e., $k = 2^1 k*$, then the truncation noise harmonics are multiples of $2^1 f_0$. This is a consequence of N, a power of 2. The limiting case occurs for $k = 2^a$, some power of 2. In that case, the truncation harmonics are harmonics of the generated frequency. These cases are demonstrated for 16 equispaced samples around a unit circle (Fig. 15). In general, for N not a power of 2 the noise period is NT divided by any common factor between k and N. Each sample has a truncation error, ϵ_i, associated with it and each line (bottom of Fig. 15) represents the sequence of sample errors generated as index k increases. Lines 1, 2, 4 are cases of a power of 2 × lowest frequency generated. Notice that the error period is the same as the generated period. Lines 3, 5, 7 represent noise periods of 16T, the longest possible because frequency index k has no factors of 2. Line 6 shows one factor of 2. As a consequence of the particular arithmetic implementation the error waveform contains only odd harmonics, i.e., it has the property $\epsilon(n) = -\epsilon[n + (P/2)]$ where P is the period.

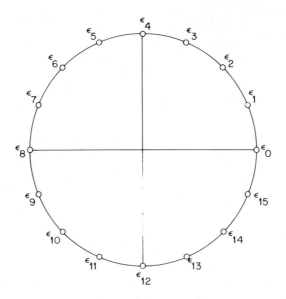

	Frequency		Noise Period
(1) Lowest	1/16T	Noise period same as generated period	= 16T
(2)	1/8T	Noise period same as generated period	= 8T
(3) •	3/16T	0 3 6 9 12 15 2 5 8 11 14 1 4 7 10 13	= 16T
(4) •	T/4	Noise period same as generated period	= 4T
(5) •	5/16T	0 5 10 15 4 9 14 3 8 13 2 7 12 1 6 11	= 16T
(6)	6/16T	0 6 12 2 8 14 1 10 0	= 8T
(7) Highest	7/16T	0 7 14 5 12 3 10 1 8 15 6 13 4 11 2 9	= 16T

Fig. 15. Truncation harmonic analysis.

Fig. 16. Error waveform, N = 16.

To bound the noise contributed by the truncation error, consider a generated frequency that is an odd factor k times $2\pi/NT$. In this case, the error waveform is of period NT or N samples long (Fig. 16). A Parseval's relation for a discrete Fourier transform over N samples can be written as

$$\sum_{i=0}^{N-1} |F_i|^2 = N \sum_{n=0}^{N-1} \epsilon_n^2$$

where F_i is a frequency amplitude defined by

$$F_i = \sum_{\ell=0}^{N-1} \epsilon_\ell \exp[-j \frac{2\pi}{N} i\ell]$$

and ϵ_n is the amplitude of the error waveform at a particular sample. The ϵ_n is a pseudorandom variable distributed uniformly over the interval -2^{-11} to $+2^{-11}$ for design 1, or -2^{-9} to $+2^{-9}$ for design 2 because of the truncation. All of the error energy in one frequency is assumed to obtain a bound. For a particular

$$|F|^2 = N \sum_{n=0}^{N-1} \epsilon_n^2$$

and assuming worst-case conditions on ϵ_n

$$|F|^2 = N^2 (2^{-11})^2 \quad \text{or} \quad |F| = N 2^{-11} \quad .$$

The desired generated frequency has an amplitude of N in the discrete transform, so that the ratio of noise amplitude to signal amplitude is 2^{-11}, the case referred to earlier. In general, representing the sample as p bits + sign, the N/S = 2^{-p}, or $-6p$ in dB. A more realistic "bound" for the cases when the noise period includes many samples can be written as

$$\sum_{i=0}^{N-1} |F_i|^2 = N^2 \left(1/N \sum_{N=0}^{N-1} \epsilon_n^2 \right)$$

where the quantity in parentheses is a sample error variance. Since the variance is $[(2^{-11}/3)]$ for uniformly distributed noise for design 1, expectedly,

$$\sum_{i=0}^{N-1} |F_i|^2 = N^2 \frac{(2^{-11})^2}{3} \quad .$$

So the bound obtained, assuming all the energy is in one harmonic, is $2^{-11}/\sqrt{3}$ noise-amplitude-to-signal-amplitude ratio or about -71 dB. Assuming that the noise waveform is white in one period, and using the fact that it contains only odd harmonics,

$$\frac{N}{2} |F|^2 = N^2 \frac{(2^{-11})^2}{3} \quad ,$$

so that the noise signal ratio = $\sqrt{2/3N} \, 2^{-11}$ for design 1, or $\sqrt{2/3N} \, 2^{-p}$ in general, which is very small for large N.

3. D/A Converter Noise

Even in the case of perfectly calculated samples driving the D/A converter, output analog samples are corrupted by noise from switching time disparities between bits and ON/OFF switching. These so-called "glitches" that occur between sample outputs depend on the initial and final words on which the converter is acting. A transition that changes many bits, such as 10000 to 01111, tends to produce more noise than a transition of only a few bits. This noise has the same periodicity as truncation noise since it depends on transitions that are periodic, if the sample train is periodic. However, all harmonics of the basic frequency are present since the "glitches" do not have odd symmetry. Assuming a "glitch" amplitude can be as high as one half of full scale out of the converter, then such a noise pulse of width Δ occurring even only once in each period will produce a noise-to-signal (N/S) rate of approximately Δ/NT. For the higher generated frequencies, such a ratio can be large. For example, if $N = 8$ and a N/S ratio of 10^{-3} (60 dB) is to be maintained, the "glitch" duration must be about 8 nsec or less, — a difficult requirement to achieve even with current techniques. Basically then, the time of nonlinearity must be small compared to the sample interval for low noise effects.

In design 1, a 12-bit converter with settling time of about 300 nsec in an interval of 1.22 μsec produced a distortion curve (Fig. 17). (Presumably, "glitches" of any consequence occupy only a small portion of this overall transient.) Since this curve

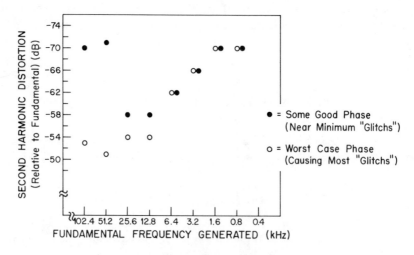

Fig. 17. Distortion curve for design 1.

represents only second harmonic distortion, it is not a result of truncation, but strictly converter distortion. Higher harmonics were smaller or equal to those shown on the curve. For generated frequencies that were not powers of 2 as given on the graph, the distortion products were less, falling between the two cases given. The larger disparity between "good" and "bad" phase of the higher generated frequencies is a consequence of the small number of samples per period so that little averaging takes place. Either a good set of samples does or does not occur so that the noise is high or low.

Even for odd-generated frequencies the predominant noise tends to be harmonics of the generated frequency rather than harmonics of the lowest frequency. This indicates that certain more significant bits of the D/A converter are delayed more or less than others, and differences between ON/OFF switch times are significant. To reduce these transition effects requires gating or sampling devices on the output of the converter, because the sampling time is small. However, for times of less than 1 μsec and linearities extending into the greater-than-60-dB range, such circuits require careful design. Using such a gate for shorting the converter output to ground during transition time, design 1 produced a worst-case harmonic of 55 dB, an improvement on the curve for the D/A converter alone.

4. Smoothing Filter Time and Frequency Response

As mentioned earlier, a final smoothing filter is needed to interpolate computed sinusoidal samples or smooth the sampled spectrum. Since this smoothing filter is the sole energy storage device in the synthesizer, the problem of time response arises. A change in frequency word input shows up in the index accumulator no later than one sampling period after the change so that the next sampling time computes a changed frequency sample and the filter response is the only time response left. As seen previously, a filter is needed to pass the band up to 204.8 kHz and reject the band above 614.4 kHz with a transition region in between [Fig. 5(b)]. The first tradeoff encountered is between rejection or attenuation at 614.4 kHz and the filter's time response. Consider a lowpass filter of 5th order (5 poles) and examine its attenuation at 614.4 kHz as well as the time response. Attenuation and step response for several filters are compared (Table 1). Sample-and-hold response adds 10 dB to these figures, i.e., if the filter input is a staircase signal rather than a narrow pulse, a frequency response with a first zero at 1/T is associated with this staircase, which is down by about 10 dB at 3/4T.

The filters (Table 1) are 1.0 dB down at 204.8 kHz and the step response is measured from 10 percent of final value to a ±10 percent window around final value. In other words, overshoots must settle down to within a ±10 percent window.

Table 1. Filter Attenuation and Step Response at 614.4 kHz

Filter	Attenuation (dB)	Step Response (μ sec)
Bessel (max flat phase)	~10	~1
Butterworth (max flat amp)	~48	~4
Chebychev (1.0-dB ripple)	~65	~7
Cauer (elliptic function filter, 1.0-dB ripple)	~85	~9

A similar trade can be effected between passband width and step response for a fixed attenuation at 614.4 kHz. If a certain attenuation is desired at that frequency, a Bessel filter has a smaller passband than does a Butterworth, and so on down the list. For a fixed order of filter, and a fixed attenuation to be achieved at 614.4 kHz, the widest passband is obtained by using the sharpest filter and this, in turn, has the poorest response time.

The response times used have been step responses to DC rather than steps of a sinusoidal input. It is clear that the time response of a step transition from one frequency to another consists of the response to two frequency steps; one to cancel the original frequency, the other to start the new frequency. The amplitude response to each of these frequency steps reflects the natural frequencies of the driven filter. Therefore, the DC step response is still a measure of the time for amplitude response to settle down.

However, it is often the instantaneous frequency out of the synthesizer that is of importance and the time this measure takes to settle down to some meaningful value. Obviously, the instantaneous frequency must be influenced by the natural frequencies of the smoothing filter and must become some steady state value after the filter response time.

If an exact response is necessary for the instantaneous frequency, careful computation is required. If a rough response time is adequate, Table 1 values are sufficient. It is true that the smaller the frequency change, the smaller the frequency and amplitude perturbation. In general, switching response is measured in the system in which the synthesizer is working.

It is also possible to reduce switching effects in both amplitude and phase response while still using a sharp cutoff filter such as an elliptic filter. This is done by extending the cutoff, and therefore, its poor phase response into the transition region where no synthesizer outputs occur. In this way, poor filter properties are negated while still taking some advantage of the sharp cutoff. A number of filters exist to satisfy smoothing requirements for any given set of amplitude or phase criteria.

F. EXTENDING THE BASIC FREQUENCY RANGE

Since the basic digital synthesizer cannot produce radio frequencies directly because of D/A speed limitation rather than logic problems, other methods must be used to produce an RF synthesizer output.

Single sideband modulation, mentioned earlier, is useful if quadrature outputs and a quadrature carrier are available (Fig. 18). The sign of one of the synthesizer outputs must change to move from upper to lower sideband, which is easily done. However, since this technique requires balanced subtraction at the output, it is difficult to achieve high suppression of the opposite side frequency of more than 40-50 dB over a wide range.

A second approach to modulating synthesizer outputs involves a single synthesizer output rather than quadrature. If the device need not produce quadrature outputs, it can work at a higher sampling or computation rate and produce a wider band of output

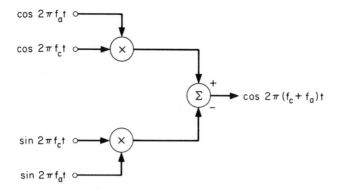

Fig. 18. Single sideband bandwidth doubling and frequency translation.

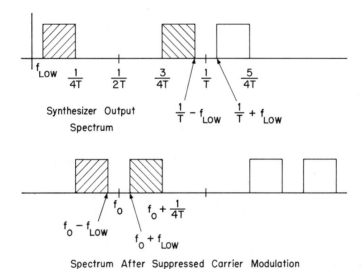

Fig. 19. Modulating single synthesizer output.

frequencies ($\sim 1/4T$). Then the lower generated frequency is limited to produce a band from $1/4T$ down to some f_{low} that requires a modest filtering after modulation (Fig. 19).

A third approach implemented at the digital level consists of computing quadrature outputs, but at different sampling times (same sampling interval) to cancel an upper or lower side frequency around some sampling harmonic (Fig. 20), i.e., if the cosine

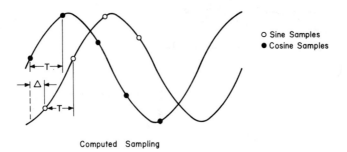

Computed Sampling

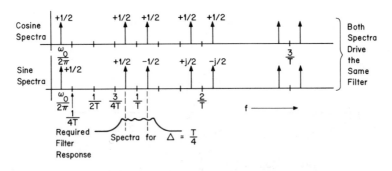

Fig. 20. Modulating synthesizer output at the digital level.

is computed at sampling times nT, then the sine must be computed at sampling times nT + Δ. To cancel the upper side frequency around the nth sampling harmonic n2π/T, Δ must be equal to T/4n, and the two computed samples are summed into the same smoothing filter (at their respective times). If the lower side frequency is eliminated, the difference between the two sample trains drives the smoothing filter. This technique follows from the expressions for the spectra of sampled cosine computed at nT and sampled sine computed at nT + Δ:

$$\text{Cosine spectrum} = \frac{1}{2} \sum_{n=-\infty}^{\infty} [\delta(\omega - \omega o - \frac{n2\pi}{T}) + \delta(\omega - \omega o - \frac{n2\pi}{T})]$$

$$\text{Sine spectrum} = \frac{1}{2_j} \sum_{n=-\infty}^{\infty} [\delta(\omega - \omega o - \frac{n2\pi}{T} - \delta(\omega + \omega o - \frac{n2\pi}{T})]$$

$$\times \exp[-j\,\frac{n2\pi\Delta}{T}] \quad .$$

This technique depends on generating narrow enough pulses out of the D/A converter to produce energy at some n2π/T sampling harmonic.

G. PHASE CONTROL AND OTHER APPLICATIONS

The phase of the synthesizer output frequency at any computation time depends on the stored value in the accumulator (Fig. 4) (ignoring any phase distortions introduced by the output smoothing filter). This permits considerable phase control. In normal operation, a frequency change is made by changing the input frequency word, k, and leaving the previous accumulator value unchanged. In this way, the new frequency is produced with no phase discontinuity since samples of the new frequency sinusoid include the last sample of the old frequency sinusoid. If, on the other hand, an arbitrary phase is desired at each frequency change, a new frequency word, k, and a new initial state for the accumulator value results. An example of such frequency switching is resetting the accumulator (and phase) to zero whenever a new frequency is generated.

The ease with which phase can be controlled suggests many applications. The ability to frequency hop under phase control permits implementation of new kinds of coherent communication systems, if channel phase is well behaved.

The synthesizer in the phase continuous mode can be used as a very linear frequency modulator. The modulating signal drives an A/D converter (unless digital already) whose output is the digital frequency control word. The use of a large encoding range and word reduces distortion and noise. If the converter is synchronous to the frequency synthesizer clock, no phase discontinuities occur, and a large deviation signal can be obtained with very high linearity.

Another class of applications is generation of various sweep signals. If the input frequency control word is incremented at certain fixed-time intervals a continuous stepped frequency waveform is generated. In the limit, this becomes a continuous frequency sweep so long as the sweep signal spectrum is not distorted by the sampling. Wired or stored frequency sequences could produce very complex sweep patterns.

The digital realization of a frequency synthesizer permits the same sample computing mechanism to generate several simultaneous frequencies. If the sampling interval is long enough (the highest frequencies needed are low enough) more than one frequency's sample can be computed. For example, by sharing the same synthesizer, samples of 10 frequencies in a 10-μsec interval can be computed at 1 μsec each if the frequencies are under 50 kHz. The change from a single synthesizer is in the accumulating mechanism and sample outputting (Fig. 21). The shift register stores the 10 frequency control words and an accumulated value for each of the 10 words. The D/A output is multiplexed synchronously to 10 output filters.

Finally, the algorithm itself, using subtables and multiplies, may be used in certain computational environments to produce digital samples of sine and cosine or complex coefficients for discrete Fourier transforms.

H. BOUND ON HIGHEST FREQUENCIES, DISTORTION AND POWER

A basic computation cycle or sampling interval will probably take 50 nsec even if direct table look-up is used with no multiplication for Schottky TTL implementation of the basic digital synthesizer design. Power dissiptation associated with such a design

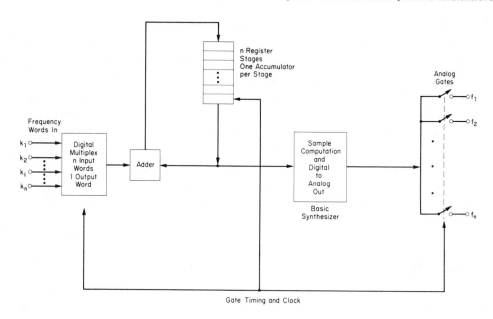

Fig. 21. Digitized frequency synthesizer facilitates generation of several frequencies simultaneously.

is in the range of 5 to 10 W, and total distortion or noise in any 100-Hz band away from the generated frequency is about -60 dB if a suitable analog gate is used to reduce D/A noise. A design implemented in complementary MOS logic and memory could probably operate at a 2-3 μsec computation cycle or sampling interval if high frequency requirements are less stringent. This small interval produces a high frequency of about 200 kHz, and will probably do so at under 1-W dissipation. Working at these longer sampling intervals, noise and distortion can probably be kept to -80 dB. Finally, implementing the digital frequency synthesizer in an emitter coupled logic (ECL) family and memory would probably reduce the sampling interval to under 30 nsec implying a high frequency of about 12 MHz. Power dissipation would probably be around 40 W. More importantly, distortion would become a severe problem as D/A converter response is severely limited at these sampling rates, and even analog gates are hard put to operate linearly at these small times. Less than -40 dB distortion is difficult to achieve with existing techniques.

REFERENCES

1. V. E. Van Duzer, "A 0-50 MHz Frequency Synthesizer with Excellent Stability, Fast Switching and Fine Resolution," Hewlett-Packard J. 15, 1-8 (May 1964).

2. A. Noyes, Jr., "Coherent Decade Frequency Synthesizers," Experimenter 38, 9 (September 1964).

3. E. Renschler and B. Welling, "An Integrated Circuit Phase-Locked Loop Digital Frequency Synthesizer," Motorola Semiconductor Products, Inc., Application Note 463.

4. G. C. Gillette, "The Digiphase Synthesizer," Frequency Technology 25-29 (August 1969).

5. J. Noordanus, "Frequency Synthesizers — A Survey of Techniques," IEEE Trans. Comm. Tech. COM-17, 257-271 (April 1969).

6. B. Gold and C. M. Rader, Digital Processing of Signals (McGraw-Hill, New York, 1969).

7. L. Jackson, "An Analysis of Limit Cycles Due to Multiplication Rounding in Recursive Digital Filters," Proc. 7th Allerton Conf., Circuit and System Theory, 69-78 (1969).

8. A. W. Crooke, "A Flexible Digital Waveform Generator for Use in Matched Filtering Applications," Arden House Workshop (January 1970).

9. H. Salinger, "Transients in Frequency Modulation," Proc. IRE 378-383 (August 1942).

10. J. Tierney, C. M. Rader and B. Gold, "A Digital Frequency Synthesizer," IEEE Trans. Audio Electroacoust. AU-19, 43-57 (March 1971).

Chapter VI
Hybrid Configurations and Frequency Stability
G. H. Lohrer

Previous chapters cover in detail three specific synthesizer systems. Performance and cost factors are presented so that a choice can be made for a certain application. Quite often, all requirements can be satisfied by one of the basic approaches and a "monolithic" design; a synthesizer using only one approach results. In other instances, a "hybrid" produces a more optimum result, e.g., a "mixed" system using two or more of the basic approaches (direct, phase-lock, digital table look-up), or a design adding circuits for other than basic frequency control.

This chapter discusses three hybrid systems: beat-frequency, drift-canceled, and system M, a direct system with a reduced number of reference frequencies. Although phase noise is probably one of the most important specifications for synthesizers, its characterization and measurement are not always straightforward. Various phase noise approaches are covered in a section on "Short-Term Stability." A brief section describes the characteristics of frequency standards that determine the long-term stability of synthesizers. The flicker-of-phase-noise-generated-by-PN-junction phenomenon that limits achievable performance, according to contemporary technology, is discussed in the last section.

A. BEAT-FREQUENCY SYSTEMS

Most basic synthesizer techniques provide moderate bandwidth; the "information," so to speak, is present on a carrier. Where narrow output bandwidth is required, the system is designed so that the specific bandwidth can be covered directly. A beat-frequency system is commonly used where a bandwidth of several octaves is needed. In the simplest form, a mixer merely "removes" the carrier by down-conversion. For example, a 5.0 to 5.1-MHz iterative output can be used to obtain 0 − 100 kHz after subtracting 5.0 MHz. Where the relative information bandwidth is narrow, wider bandwidths can be obtained (at the expense of spurious/noise degradation and loss of resolution) by multiplication before the mix. The 5.0 and 5.1 minus 5.0 MHz now appear as 50 to 51 minus 50.0 MHz to yield a 0 − 1 MHz synthesized output band.

Whereas multiplication produces wider output bandwidths, division can increase the absolute resolution of a synthesized output band (at the expense of the bandwidth covered). Unlike multiplication, which generally has to be accomplished for both inputs ahead of the output mixer (narrowband), digital dividers can follow the mixer and process the output. Once a certain bandwidth and resolution have been obtained, the same relative resolution at lower and lower output frequencies can be maintained by this technique.

All iterative systems permit the addition of resolution at the least significant digit or "low" end. In view of this, it would appear that a wide bandwidth in the repetitive processing channel is advantageous to obtain without multiplication the widest possible output coverage as well as fast frequency switching. This is an economic

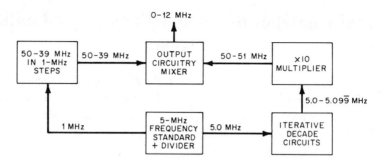

Fig. 1. Modified beat-frequency system with frequency-
variable inputs.

tradeoff. Bandwidths of 10 percent are seldom exceeded because of requirements to filter unwanted products. This leads to a 100-MHz carrier for a 10-MHz output bandwidth. If this is the output of a decimal iterative system after division, then the divider input is 1 GHz. As some frequencies are not handled economically in dividers, the iterative approach is not usually carried to the most significant digit, but a modification to the simple beat-frequency system can be used in which both inputs are made frequency variable. In an example of such a system (Fig. 1) the multiplication has been retained, but the previously fixed 50.0-MHz signal is changed to vary from 50.0 down to 39.0 MHz in 1-MHz steps. How far this approach is carried depends on the way the most significant steps are obtained and the final mixing ratio. To avoid a 5th-order spurious signal in the output, $(f_o/f_x)^*$ has to remain above 0.75, preferably with some margin for a filter with finite cutoff. In this example with a 12-MHz output frequency, the 5th-order product appears at 15 MHz.

This brings us to an important consideration. To avoid output mixer spurs lower than 7th order in a 0 to 100-MHz beat-frequency system, it is necessary to beat 400-500 MHz and 500 MHz. Obviously, those "extra" 400 MHz are there only to improve the final mix ratio. This is not an economic arrangement, and in addition, leads to less than optimum results if those two frequencies are not processed very carefully. In some way, these inputs have to be multiplied from a standard of 10 MHz (the absolute frequency is not important). With a phase jitter, $\Delta\phi$, present on the standard, two ideal multipliers present to the mixer 40 and 50 $\Delta\phi$ of the same spectral distribution with no delay. After subtraction, 10 $\Delta\phi$ of jitter is the best that can be expected for a 100-MHz output signal. With real multipliers, delays exist, and at some offset, $f = 1/2 \, \Delta\tau$, the two multiplication factors add rather than subtract so that jitter on the 100-MHz output signal is 90 $\Delta\phi$.

To minimize this effect, consider a system with identical inputs, as high as possible, to the two sides of the mixer (Fig. 2). The "common" frequency comes from a drift-canceled oscillator (DCO) that need not be locked, and whose frequency drift is

---

$^*f_o$ = output frequency; $f_x = X^{th}$ order intermodulation products.

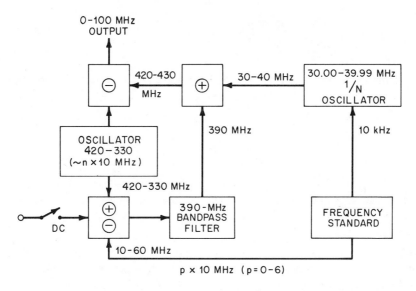

Fig. 2. 100-MHz synthesizer with drift-canceled oscillator.

self-canceled in the output. This 100-MHz synthesizer with 10-kHz resolution uses essentially two-digit modules. The fine resolution module (10-kHz steps) phase locks a 30- to 40-MHz oscillator to a 10-kHz reference through a programmable divider of n = 3,000-3,999. The effect of a coarse resolution module (10-kHz steps) is achieved by the DCO, 390-MHz bandpass filter (BPF), and associated mixers in the following fashion. The DCO is tuned approximately to a multiple of 10 MHz within the 330-to 420-MHz band. Its output is mixed with the first five harmonics of 10 MHz (as well as DC) in the lower mixer to produce a line close to 390 MHz, which is passed by the BPF. (The other lines are rejected by this filter.) This 390-MHz line is mixed with the output of the fine resolution module to provide one of the two inputs to the output mixer. The DCO provides the other input to the output mixer.

To understand the drift cancellation, suppose that the DCO output is at (410 + Δf) MHz, where Δf is less than one half the bandwidth of the 390-MHz BPF. Then the line passed by the BPF will be at (390 + Δf) MHz. If the fine resolution module outputs 34.510 MHz, the right-hand input to the output mixer is then (424.510 + Δf) MHz. Thus the output frequency obtained by mixing with the DCO output is (424.510 + Δf) − (410 + Δf) = 14.510 MHz. The error, Δf, in the DCO frequency is canceled at the output.

To get the desired cancellation, the same delay, τ, must exist in both paths to the final mixer. With finite bandwidth, cancellation is, of course, not perfect at all frequencies. The cancellation voltage ratio can be approximated by

$$ C = \frac{\Delta \phi_{out}}{\Delta \phi_{in}} = 2\pi f_m \Delta \tau $$

where f_m is the modulation frequency (noise) and $\Delta \tau$ is the differential delay in the branches from the oscillator to the final mixer, as long as $C \ll 1$. The 390-MHz filter is potentially the main contributor and its delay is

$$\tau = \frac{\text{Number of poles}}{2\pi \text{ bandwidth}} \quad .$$

Neglecting other delays for the moment (using τ_{filter} as $\Delta\tau$) and applying some numbers to get a feel for the magnitudes, assume that a two-pole, 5-MHz filter is used. At 10 kHz off the carrier:

$$C_{10 \text{ kHz}} = \frac{2\pi \cdot 10^4 \text{ Hz} \cdot 2}{2\pi \cdot 5 \cdot 10^6 \text{ Hz}} = 4 \times 10^{-3} = -48 \text{ dB} \quad .$$

This is the factor by which the free-running oscillator's own noise spectrum is improved at a 10-kHz offset frequency. The oscillator's contribution, determined by oscillator quality and cancellation (control of differential delay), should be below that of the frequency standard.

Another application for the DCO or "Wadley System" is the so-called "Triple-Mix" system (Fig. 3) that selects one component of a "comb" of equidistantly spaced frequencies by two fixed filters and two associated mixers. Input from the next lower

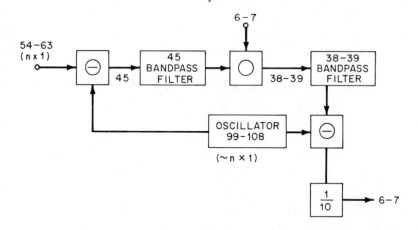

Fig. 3. Triple-mix system. (All frequencies in MHz.)

stage in a repetitive iterative system may be accepted by the third mixer. The output divider reduces the significance of the previous stage output and the DCO by the base factor. The oscillator can be (1) step-tuned, (2) continuously tuned, or (3) separate oscillators can be used. The fixed-filter bandwidth is a compromise between attenuation of comb pickets and noise cancellation of the oscillator around the loop.

B. SYSTEM M

In an iterative decimal system, input from the preceding lower-order stage and the selected digit frequency from a comb of ten frequencies are mixed and divided by ten. This produces steps at the output of the stage that have 1/10 the frequency separation of the comb and reduces digits from previous stages by a factor of 10 in significance. Frequently, division by 10 is not made in a single step, but in two dividers with ratios of 2 and 5. If pickets of normal spacing are introduced after the divide-by-2 stage, their "weight" is 2, compared to pickets that have passed through the divide-by-2 stage. With a choice of weight, 1 or 2, 10 digits with only four picket frequencies can be produced.

Can the iterative stage be designed to use two identical sets of comb frequencies? The general equation governing iterative stages is (Fig. 4)

$$[\frac{F}{k} + \frac{F(k-1)}{k}] \frac{1}{k} = \frac{F}{k} \quad .$$

Consider F the carrier or band-end frequencies.

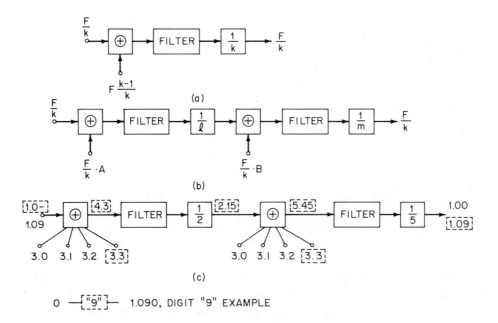

Fig. 4. Iterative stages (a) general, 1 divider, (b) general, 2 dividers, (c) decimal.

Splitting up the division into ℓ and m, where $k = \ell \cdot m$ and adding a mixer, but leaving the input/output conditions unchanged, arrangement Fig. 4(b) is obtained. The general equation is

$$[\frac{F}{k} (A + 1) \frac{1}{k} + \frac{F}{k} \cdot B] \cdot \frac{1}{m} = \frac{F}{k} \quad .$$

Simplifying gives the condition:

$$k - 1 = A + B \quad .$$

Let A = B, then $k - 1 = B(\ell + 1)$ for k = 10, ℓ = 2, and A = B = 3. A numeric example with F (bottom frequency) of 1 MHz, and output steps of the first decade of

10 kHz is shown in Fig. 4(c). "Weight one" steps (with a division of 10) are obtained for 100-kHz input pickets. From 3.0, 3.1, 3.2 and 3.3 MHz, output steps of 0, 10, 20 and 30 kHz are obtained; the same pickets fed into the second mixer produce "weight two" digits of 0, 20, 40, 60 kHz so that by proper addition of these outputs all 10 steps from 0-90 kHz are obtained.

C. SHORT-TERM STABILITY

Short-term stability, spectral noise distribution, signal-to-noise ratio (SNR), and fractional frequency deviation describe disturbances on a signal that ideally should be of single frequency, but typically, is a carrier with sidebands. Phase modulation (PM) sidebands are of particular concern for synthesizers because amplitude modulation (AM) sidebands have usually been removed by limiting. PM, however, can also result from the conversion of a single, low-level, spurious frequency into a pair of PM sidebands. (A single sideband can be thought of as the superposition of two pairs of sidebands of equal amplitude and frequency, one AM and one PM. In the limiter, the AM is removed, and the pair of PM sidebands — each of 6 dB lower than the original single spurious frequency — remain.)

Angle modulation is described by the equation:

$$V(t) = A \cdot \sin(\Omega_c t + \Delta\phi(t))$$

where $\Omega_c = 2\pi F_{carrier}$ and $\Delta\phi(t)$ is the noise modulation. This expression can be expanded to:

$$V(t) = A \sin\Omega_c t + \cos\Delta\phi(t) + A \cos\Omega_c t \sin\Delta\phi(t)$$

which for the normal case of small phase noise becomes:

$$V(t) \simeq A \sin\Omega_c t + A\Delta\phi(t) \cos\Omega_c t \quad .$$

Thus the signal V(t) consists of an unmodulated "carrier" $A \sin\Omega_c t$ and a small AM signal $A\Delta\phi(t) \cos\Omega_c t$ in phase quadrature.

One of the simpler ways to specify this noise modulation on a carrier is the so-called carrier-to-phase noise ratio. A measurement technique is illustrated in Fig. 5. To detect noise modulation in a phase detector, a reference of the same frequency and a 10-20 dB better SNR is needed. Frequently, this is not available and two signals of the same "impurity" have to be compared. Their noise has to be uncorrelated; then the result can simply be reduced by 3 dB. The measurement relates two output levels obtained from the phase detector. One is the integrated, detected noise inside the bandwidth of the lowpass filter; the other, the reference level for 1 rad $\Delta\phi$. A simple way to establish the reference is to use a beat, where one signal is offset slightly. If very accurate results are wanted, it is better to calibrate by a known small phase deviation, say $\Delta\phi = 10$ mrad, and add 40 dB to the measurement:

$$\frac{S}{N_\phi} = 20 \log \frac{E_{1\ rad}}{E_{noise}} \quad .$$

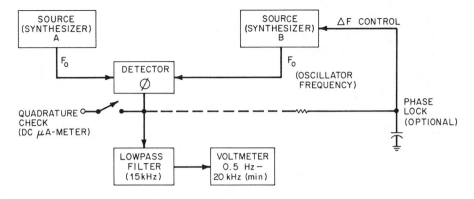

Fig. 5. Carrier/phase noise ratio (broadband).

Typical results for synthesizers range from 40-70 dB for a 0.5 Hz to 15-kHz measurement bandwidth.

Currently, the most popular method for characterizing synthesizer phase noise is the spectral plot of the foregoing broadband measurement in terms of the carrier-to-single-sideband ratio (Fig. 6). This setup is very similar to Fig. 5 except that instead of a broadband integrating voltmeter, a selective wave analyzer, preferably with a 1-Hz bandwidth, is used. Carrier level reference is obtained by a small offset, or again, by establishing the voltage for 1 rad (peak). The wave analyzer readings of demodulated noise have to be reduced by 6 dB for the single-sideband (i.e., one-sided) plot. Preamplification may be required for demanding applications, and attention to noise figure is important. A phase-lock loop may be used to maintain quadrature of the two signals at the phase detector, if they are not highly stable.

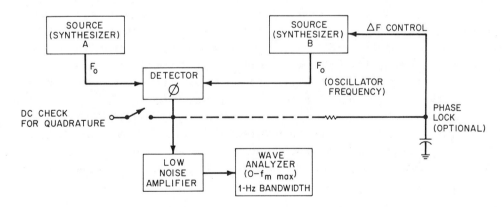

Fig. 6. Carrier/phase noise ratio (spectral distribution).

Within the bandwidth of that loop, the signals are then no longer uncorrelated. The cutoff of the loop, therefore, should be well below the lowest frequency, f_m, of interest (Fig. 7).

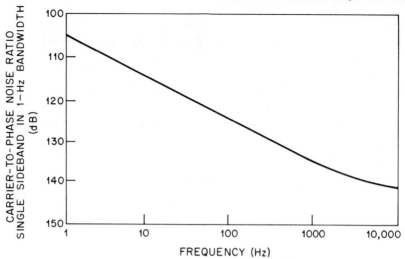

Fig. 7. Spectral distribution of phase noise for a typical synthesizer output.

In some applications, frequency deviation is used as a measure of purity of the output signal. In this case, it is assumed that $\Delta\phi(t)$ can be modeled as

$$\Delta\phi(t) = \Delta\phi \, \sin \omega_m t$$

where $\omega_m = 2\pi f_{modulation}$. Then the instantaneous frequency is

$$F_i = F_c + \Delta F$$

and the deviation is $\Delta F = f_m \, \Delta\phi \, \cos \omega_m t$.

If the equation for $V(t)$ is expanded, the familiar infinite sideband distribution with spacing f_m is obtained in which carrier and sideband amplitudes are given by the Bessel functions of orders 0, 1, 2, etc. For values of $\Delta\phi \ll 1$ the ratio of first sideband to carrier amplitude, $J_1/J_0 = \Delta\phi/2$, and higher order J's approach 0.

A good frequency discriminator is needed to measure this deviation, and it may be necessary to first process the signal by heterodyning (only with a signal of higher purity and in a low noise mixer) or by multiplying for greater sensitivity. (Deviation is increased directly by the multiplication factor.) Again, a broadband indicator may be used, or a narrowband spectral distribution of the deviation may be established. To relate this measurement to the single-sideband phase noise plot, consider some numbers for a narrowband case. If it is determined that a ±0.1-Hz peak deviation was due to noise at 1 kHz off the carrier measured in a 1-Hz bandwidth, then

$$\Delta\phi = \frac{\Delta F}{f_m} = \frac{0.1 \text{ Hz}}{10^3 \text{ Hz}} = 10^{-4} \quad .$$

The sidebands belonging to this $\Delta\phi$ are 86 dB down from the carrier and no adjustment for bandwidth is needed as both measurements were made with 1-Hz bandwidth.

The third method produces the result in the form of a fractional frequency deviation, $\Delta F/F_o$. Simply stated, it sets forth the expected accuracy of the frequency under discussion for a particular observation time. The most accepted measure of this instability is the one signal value of the "Allan variance" for two consecutive ($N = 2$) measurements of frequency:

$$\sigma(2, T, \tau) = \frac{1}{F_o} \sqrt{\frac{1}{2N} \sum_{1}^{N} (F_i - F_{i-1})^2} \quad .$$

The measurement is executed by measuring N samples of the frequency (or period) of duration or averaging time, τ, separated by dead time, T. In principle, a counter can perform this simple, straightforward measurement. When plotted, the resulting N samples form a histogram clustered around the nominal f_o with normal distribution. There are, however, a number of difficulties if the frequency to be evaluated is spectrally most pure. In a direct measurement, the counter time base has to be better than the signal under test. When this is not the case, this requirement can be eliminated by comparing two of the (uncorrelated) signals to be measured, and magnifying the fractional instability by down-conversion to the audio range, where single- or multiple-period measurements yield averaging times, τ, of the proper magnitude.

The second difficulty may arise from input noise in the counter. A 40-dB SNR permits only a ±0.3 percent single-period measurement. A low-noise preamplifier may have to be used to improve the effective SNR at the counter input. There are

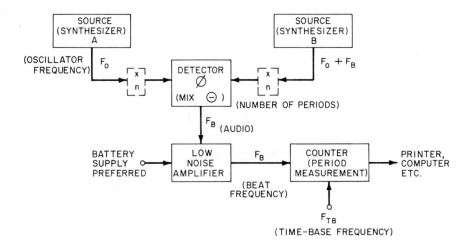

Fig. 8. $\Delta F/F$, period measurement.

numerous other pitfalls such as ground loops and close-in discrete spurious frequencies that make the measurement difficult. These problems are illustrated in Fig. 8, where

$$F_B = \text{beat frequency} \qquad \tau = \text{averaging time}$$

$$F_{TB} = \text{time-base frequency} \qquad R = \text{resolution}$$

$$N = \text{number of periods} \qquad F_o = \text{oscillator frequency.}$$

Resolution of the system consists of two parts:

1. Resolution of the counter in the period mode:

$$R_{C1} = \frac{F_B}{F_{TB}} \text{ (single period)}$$

$$R_{C2} = \frac{F_B}{F_{TB} \cdot M} \text{ (M, periods, multiple)} \qquad .$$

2. Expansion term due to mixing:

$$E = \frac{F_B}{F_o} \qquad .$$

Together the resolution becomes:

$$R_{total} = R_{C2} \cdot E = F_o \cdot \frac{F_B^2}{F_{TB}} \cdot M$$

with

$$\tau = \frac{M}{F_B} \qquad ,$$

$$R_T = \frac{M}{\tau^2 \cdot F_o \cdot F_{TB}} \qquad .$$

For example, if:

$$F_o = 100 \text{ MHz}; \; F_B = 1 \text{ kHz}; \; M = 100;$$

$$\tau = 0.1 \text{ sec}; \; F_{TB} = 10 \text{ MHz} \qquad ,$$

then

$$R_T = \frac{100 \text{ sec}}{10^{-2} \text{ sec}^2 \; 10^8 \; 10^7} = \frac{10^2}{10^{13}} = 1 \times 10^{-11} \qquad .$$

D.   REFERENCE FREQUENCY CONSIDERATIONS

The synthesis circuits described operate on the output of a reference generator or frequency standard. The synthesizer proper then multiplies the standard's frequency by a rational number. All of the systems discussed — direct synthesis, phase-locked loops, digital table look-up — are coherent systems, meaning that they use only a single reference input rather than combinations of various stable sources.

In a coherent system, the relative long-term stability of any output frequency reflects directly the long-term stability of the reference, as expected, since the multiplication or synthesis factor selected involved only certain fixed arithmetic operations; and the nature of the processing circuits is such that they cannot, in the long term, add or subtract units of frequency. To provide a given long-term stability at the synthesizer output, a standard from one of the following broad categories can be selected:

1. Room-temperature crystal oscillators that age at a few parts per million monthly; temperature coefficients of 1 to 5×10^{-7}/°C. Improvement in temperature behavior can be made over limited temperature ranges by electrical or mechanical compensation. Generally, these crystals are AT cuts, fundamental mode. (AT cuts denote a particular orientation of the crystal blank with respect to the axis of the crystal from which they are cut.)

2. Oven-type crystal oscillators, mainly with proportional control. Aging from 1×10^{-8} to 10^{-10}/day. Temperature coefficients of 1×10^{-10}/°C, or better. These oscillators use AT cuts, both fundamental and 5th overtone.

3. Atomic standards: Rubidium gas cell, cesium beam, hydrogen maser. Long-term stabilities are 10^{-11} to 10^{-13}, and temperature coefficients usually fractions thereof. Typically, these units have high-grade, oven-type, crystal oscillators that are slaved, sometimes with a selectable time constant, to the atomic resonator.

If the frequency stability concept is applied to short-term stability, averaging times are one second and less. The choice made regarding long-term stability of the reference also affects short-term performance. To an extent, one improves at the expense of the other. Fifth-overtone crystal oscillators with the best long-term behavior operate at drive levels of 1 mW or less. Their short-term stability is close to that set by the thermal noise level of the crystal (its equivalent series resistance). Higher drive levels that improve the SNR, affect medium- and long-term stability adversely. Relative frequency stability vs averaging time is plotted in Fig. 9.

The short-term stability of the complete synthesizer contains contributions from the reference source and frequency processing circuits, the synthesizer proper. Higher drive levels improve the short-term stability of the reference. If the resulting long-term stability is not satisfactory, the high-level oscillator can be phase-locked with a long time constant to a low-level oscillator, which can impart its good long-term stability to the combination. A crystal filter with high drive can also be used following a low-level oscillator.

E. FLICKER-OF-PHASE NOISE

1/f noise present in semiconductors produces direct intrinsic phase modulation when an RF current is passed through a PN junction. Even though the level of 1/f

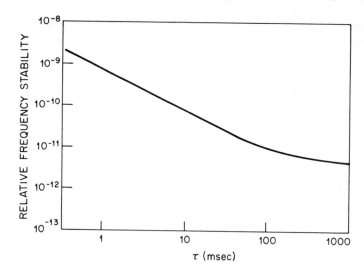

Fig. 9. ΔF/F for a typical 5-MHz, 5th-overtone, crystal oscillator.

noise may be different among various devices, the amount of PM produced seems almost constant. Following a $1/f$ (10 dB per decade) law, a one-sided carrier-to-phase noise density at 1 Hz of 115 dB/Hz is obtained. Although improvements by negative feedback are possible, this phenomenon sets a noise floor on processing of frequencies in synthesizers. When multiplication and division are involved to multiply by a fraction, the sequence of operations becomes important, if the lowest phase noise is desired. Each transistor amplifier or multiplier is a source of this flicker-of-phase noise.

A simple example is the generation of 14 MHz from a 10-MHz standard, $14 = (7/5) \times 10$; the sequence (Fig. 10), 10×7 first and $70/5$ second is preferable to the one that puts division first and multiplication second. If 2 MHz are produced first, the noise modulation in the next PN junction has to be multiplied by 7. The result is 11-14 dB worse than it has to be, if the 10-MHz standard is state-of-the-art with phase noise near the above −115 dB/Hz at 1 Hz.

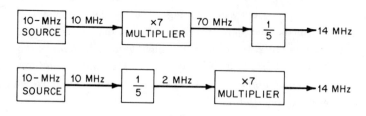

Fig. 10. 7/5 amplifier.

Similarly, to produce 34 MHz from 10 MHz $(3 \times 10) + (10/10) \times 4$ would produce the better result; $(10/10) \times 34$ should not be used.   This requires higher frequency dividers that are becoming available in digital form.   The rule — multiply first, divide after — also eases requirements on the dividers regarding the phase jitter they introduce due to noise in their input circuits.

Operating at low noise levels, a crystal oscillator has been reported that in the 1-1000 Hz range exhibits noise 10 dB better than the $1/f$ plot passing through -115 dB/Hz at 1-Hz offset.   The lowest contribution from following circuits is essential to fully realize the potential of such a standard.

REFERENCES

H. Brandenberger, E. Hadorn, B. Halford, J. H. Shoaf, "High Quality Quartz Crystal Oscillators: Frequency Domain and Time Domain Stability," Proc. 25th Ann. Freq. Cont. Symp., 226 (April 1971).

W. F. Byers, "70-MHz Synthesizer Joins the Family," General Radio Experimenter (September 1966).

M. Frerking, "Short-Term Stability Measurements," Proc. 21st Ann. Freq. Cont. Symp., 273 (April 1967).

D. Halford, A. E. Wainwright, J. A. Barnes, "Flicker Noise of Phase in RF Amplifiers and Frequency Multipliers: Characterization, Cause and Cure," Proc. 22nd Ann. Freq. Cont. Symp., 340 (April 1968).

L. Illingworth, "Digital Methods Synthesize Frequency," Electronic Design (May 23, 1968).

V. Kroupa, "Theory of Frequency Synthesis," IEEE Trans. Instr. and Meas. (March 1968).

D. B. Leeson, "Short-Term Stable Microwave Sources," Microwave J., 59 (June 1970).

G. H. Lohrer, "Faster Switching for 1160-Series Synthesizers," General Radio Experimenter (May/June 1969).

D. Martin, "Frequency Stability Measurements by Computing Counter System," Hewlett-Packard J. (November 1971).

M. Maruyama, "Major Decimal System in Frequency Synthesis," Electronics and Control J. (May 1962)

D. G. Meyer, "An Ultra-Low Noise Direct Frequency Synthesizer," Proc. 24th Ann. Freq. Cont. Symp. (April 1970).

A. Noyes, Jr., "Coherent Decade Frequency Synthesizers," General Radio Experimenter (September 1964).

F. Oropeza, J. Schoenberg, "Binary Frequency Synthesis," Frequency (September/October 1966).

R. E. Paradysz, W. L. Smith, "Measurement of FM Noise Spectra of Low-Noise VHF Crystal Controlled Oscillators," IEEE Trans. Instr. and Meas. (December 1966).

H. P. Stratemeyer, "The Stability of Standard-Frequency Oscillators," General Radio Experimenter (June 1964).

B. M. Wojciechowski, "Theory of Frequency Synthesizing Networks," Bell System Tech. J. (May 1960).

163

Subject Index

component
  amplitude modulation, 59
  crosstalk, 59
  frequencies, 49, 65
  frequency generator, 55
computation
  of sinusoidal component, 127
control
  parallel, 3
  phase, 147
  shift rate, 3
  word, 3
Craiglow, 5, 21
Crispin, 11, 21
crosstalk, 59
  mechanisms, 52
cutoff frequency, 29, 34, 61

D

Dale, 10
D/A
  conversion, 43
  converted, 121
  converter noise, 142
  effects, 138
DC
  control circuitry, 36
  error signal, 37
  errors, 105
  levels, 2, 9, 71, 73
Dean, 12, 21
Dering, 10
Digiphase
  modulation, 93
  principle, 72, 106, 117
  system, 111
  technique, 39
digital
  communication, 17
  divider, 151
  frequency synthesizer, 5, 42, 91, 121
  recursion oscillator, 121

direct
  chain synthesizer, 27
  computation, 121
  synthesis, 25, 26, 30, 37, 49, 52, 65, 151
  table look-up, 25, 39, 121
divide-by-N, 96
  counter, 32
  phase-lock synthesizer, 25, 32
divider, 26, 30, 39, 103
  digital, 151
division ratio, 27
domain
  frequency, 4, 14
  time, 4
drift-canceled
  oscillator, 26
  system, 151
Drouilhet, 16

E

East, 5, 22
electromagnetic spectrum, 12
emissions
  spurious, 3, 6
Evanzia, 6, 22
Evers, 6, 22

F

filter
  cutoff rate, 135
  smoothing, 121
  transfer function, 112
filtering synthesizer, 57
Finden, 12, 22
Flicker, 5, 22
flicker-of-phase noise, 151, 161
FM tuner
  high-fidelity, 37

phase-locked, 26
recursive limit-cycle, 122
reference, 1, 71
tunable, 31
output
frequencies, 2
noise, 138
quadrature, 127
smoothing filters, 128, 138

P

Parseval's relation, 140
Perrigo, 6, 21
Peterson, 5, 23
phase
continuous, 2
continuous mode, 147
control, 147
detector modulation, 74
difference, 13
jitter, 139, 152, 163
margin, 104
modulation, 47, 138
modulation sidebands, 156
noise, 3, 55, 57, 61, 105, 138
noise spectrum, 47
perturbations, 1
settling time, 52, 65
settling transient, 47, 49
spectrum, 47
tracking, 12
variance, 55, 57, 62
phase-locked loop, 3, 12, 25, 55, 75, 157
digitally controlled, 70, 76
oscillator, 26
synthesis technique, 38
synthesizer, 34, 37, 69, 129
point-to-point communication, 5
presetting scheme, 77, 102
pretune, 102, 109
characteristics, 109
voltage, 34

Primich, 11, 21
programmability, 2
programmable
divide-by-N, 25
divider, 32, 153
frequency, 50
phase-lock synthesizer, 36

Q

quadrature, 157
outputs, 127

R

radar
coherent Doppler, 20
frequency-agile, 19
mapping, 19
reference signal, 11
radar cross section
measurements, 11
radio
broadcasting channels, 6
frequency spectrum, 12
RASSR, 9
Raven, 20, 23
Ray, 19
read-only memory, 25, 43, 91, 123, 124, 132
storage, 124
reference, 104
frequency, 1, 7, 37, 42, 47, 71, 75, 93, 99, 112, 151
oscillator, 1, 71
signal, 10, 98
tone, 49, 55
Reliable Advanced Solid-State Radar, 9
resolution, 2, 5, 16, 18, 20, 71, 75 94, 101, 151
fine frequency, 37
resonance
frequencies, 7
Reynolds, 5

Ribour, 6, 23
Robillard, 11, 21
Rosen, 17
Rozov, 24

S

Sacha, 6, 22
sample-and-hold, 98, 100
sampled data technique, 121
sampling harmonics, 138
satellite communications, 12
Schafer, 9, 24
Searle, 21
selectable sideband upconversion, 43
settling time, 97, 112
Shields, 5, 24
short-term stability, 1, 4, 20, 156, 161
sideband
    offset frequency, 30
    suppression, 102
Siegel, 11
signal stability, 71
simultaneous frequencies, 147
single sideband, 5
    mixer, 72, 75
    modulation, 144
    plot, 157
    system, 93
    upconversion, 41
sinusoidal computation, 127
Skolnik, 18
smoothing filter, 121, 133, 146
source phase effects, 138
spectral purity, 9, 41, 43, 47, 52, 117
spurious
    emission, 3, 6
    frequencies, 60
    inband
        mixing outputs, 27, 30, 36, 58
        products, 60
        signals, 152
    noise, 3, 8, 151

output suppression, 30, 43
radiations, 2
requirements, 20, 42
sideband expression, 36
sidebands, 35
signal levels, 9
signals, 3
supression, 31, 42, 65
tone, 58
stability
    closed-loop, 112
    frequency, 4
    long-term, 12
    short-term, 1, 4, 20
Stevens, 6
stopband
    frequency region, 8
    rejection, 30
supression, 30
    noise, 106
sweep signals, 147
swept-frequency
    ionogram, 6
    output, 71
switching
    speed, 3, 17, 19, 27, 53
    time, 2, 18, 20, 28, 30, 37, 42, 47, 138
synthesis
    direct, 25, 37
synthesizer
    band, 76
    digital frequency, 5, 91
    filtering, 57
    multistage iterative direct, 25, 42
    nonfiltering, 57
    phase-lock, 38
    programmable phase-lock, 36
    properties, 1
    resolution, 16
    switching, 16
system M, 151

T

Tactical Transmission System, 16
TATS, 16, 27
Taylor expansion, 47, 61
television
   broadcasting, 7
time
   domain, 4
   smoothing filter, 143
tracking carrier, 5
transceivers, 6
   Citizen's Band, 37
transfer function, 14, 114
transient
   errors, 104
   perturbation, 104, 112
   response, 33
   testing, 43
Triple-mix system, 154
truncation, 138
   noise, 139
TTL packages, 90, 133, 138
Tveten, 6

two-synthesizer loop, 91
Tykulski, 5

U

underwater communications, 20
upconversion carrier, 43
Utton, 10

V

Van Duzer, 12, 24
variance, 4

W

Wadley system, 124
Williams, 20, 24
Wilmshurst, 10
Wood, 17
Wright, 6

Y

Young, 5, 24

Biographies

Jerzy Gorski-Popiel, editor of this book, was born in Warsaw, Poland, Feb. 7, 1933. He received the B.Sc. and Ph.D. degrees in electrical engineering from Queen Mary College, University of London in 1959 and 1967, respectively. From 1959 until 1968 he worked on active network theory and linear systems at the A.E.I. Research Laboratory in Blackheath, London, England. Since 1968 he has been a staff member at the M.I.T. Lincoln Laboratory, Lexington, Mass., where he has been involved in work on nonlinear network theory, frequency synthesizer and digital signal processing. Dr. Gorski-Popiel is the author of numerous scientific papers and holds 14 patents in active circuit design.

Thomas S. Seay was born in Birmingham, Ala., December 4, 1941. He received the S.B., S.M., and E.E. degrees in electrical engineering from the Massachusetts Institute of Technology, Cambridge, Mass., in 1964, 1965 and 1966, respectively. Since 1966 he has been involved in military satellite communications system and terminal design for the Communications Division of the M.I.T. Lincoln Laboratory, Lexington, Mass., where he is currently associate leader of the Communication Systems Group.

Ben H. Hutchinson, Jr. was born in Atlanta, Ga., in 1937. He received the B.E.E. degree from the Georgia Institute of Technology in 1959, and the M.S.E.E. and E.E. degrees from M.I.T. in 1961 and 1962, respectively. A staff member at the M.I.T. Lincoln Laboratory, Lexington, Mass., since that time, he was appointed assistant leader of the Communication Systems Group in 1971, and then associate leader of the Special Communications Technology Group. Mr. Hutchinson has worked on receiver circuitry, RF and digital modulation and signal processing systems and a wide variety of frequency synthesizers for satellite communication systems in both a design and supervisory capacity.

Carl H. Gundel is a senior engineering specialist at GTE Sylvania's Eastern Division, Needham, Mass. He has had technical design and development responsibility for a succession of high performance frequency synthesizers and related spectral processing circuitry for sophisticated communication systems. Mr. Gundel holds a B.S. in physics from Case Institute of Technology and an M.S.E.E. from Northeastern University. He is a member of Tau Beta Pi and Sigma Xi.

Joseph Tierney was born in New York, N.Y., Jan. 14, 1934. He received the S.B. and S.M. degrees in electrical engineering from the Massachusetts Institute of Technology, Cambridge, Mass., in 1956. While at M.I.T. he was a co-op student at the Bell Telephone Laboratories, and spent the year 1955-1956 as a research assistant at the M.I.T. Research Laboratory of Electronics. He has been with the M.I.T. Lincoln Laboratory, Lexington, Mass., since 1961 working in the areas of speech compression and satellite communications at both a systems and circuits level. For the past several years he has been involved in the design of digital signal processing devices.

George H. Lohrer is a graduate of Technische Hochschule (Dipl Ing EE), Karlsruhe, West Germany. His first employment was with G. Lorenz A.G. (ITT); then for Canadian GE, Philips Electronics, and the National Company in various capacities in the VHF-UHF field. In 1961 he joined General Radio, West Concord, Mass., where he has specialized in synthesizers and is currently project engineer for synthesizers. He is a senior member of the IEEE and a registered professional engineer in Ontario, Canada, and Massachusetts.
